GNVQ Advanced

Engineering Mathematics

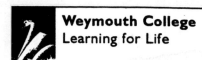

Weymouth College
Learning for Life

Alex Greer
Graham Taylor
and
Ruth Howkins
Gloucestershire College of Arts and Technology

STANLEY
THORNES

First published in 1996 by
Stanley Thornes Publishers Litd
Ellenborough House
Wellington Street
Cheltenham
GL50 1YW
UK

96 97 98 98 00 / 9 8 7 6 5 4 3 2 1

A catalogue record for this book is available from The British Library.

ISBN 0 7487 2387 0

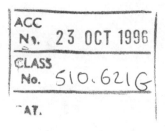
Typeset by Tech-Set Ltd, Gateshead, Tyne and Wear
Printed and bound in Great Britain at Scotprint Ltd, Musselburgh

Contents

Introduction to GNVQs in Engineering

GNVQs (General National Vocational Qualifications) are a new type of qualification available for students who want to follow a course linking traditional areas of study with the world of work. Those who qualify will be ideally placed to make the choice between progressing to higher education and applying for employment.

They can be taken at Foundation, Intermediate or Advanced level, and consist of mandatory units (those everyone must study), optional units (from which everyone must choose), additional units (which may be chosen to top up skills and knowledge) and core skills units (covering communication, IT and application of number).

GNVQ specifications and assessment

The units specifications may appear complex at first sight. However, each unit is structured in the same way:

Elements	which focus on specific aspects of the unit.
Performance criteria	which tell you what you must be able to do.
Range	which tell you what areas you must be able to apply your knowledge to.

To get your GNVQ you must pass each of the mandatory units, as well as those options you choose. Mandatory units are assessed by an end of unit test, comprising a multiple choice paper, and by the satisfactory assessment of your portfolio for that unit. Optional units are assessed by portfolio only. Further guidance is provided in Chapter 15 of this book.

You may be awarded a pass, a merit or a distinction. Guidance on how to achieve the higher grades will be available from your lecturer.

About this book

This book is written for mandatory Unit 8 of the Advanced GNVQ – *Mathematics for Engineering (Advanced)*. It is divided into three parts, corresponding to the three elements into which the unit is divided:

Element 8.1   Use number and algebra to solve engineering problems

Element 8.2   Use trigonometry to solve enginering problems

Element 8.3   Use functions and graphs to model engineering situations and solve engineering problems.

Within the three parts, each chapter closely follows the performance criteria, paying particular attention to the range, amplification and evidence indicators.

At the end of each part there is a selection of problems and an assignment, suitable for inclusion as portfolio evidence.

At the end of the book is a detailed section on portfolio preparation and the end of unit test, with useful guidance on techniques and useful hints for answering multiple-choice qustions. Finally, there are six specimen test papers, with answers. Each paper covers the full range of topics as set out by the unit test specifications for mandatory Unit 8.

Authors' note

This book has been written primarily for students taking a GNVQ Engineering Course. However, it is also well suited to the following courses:

- BTEC National Diploma in Mechanical Engineering
- BTEC National Diploma in Electrical or Electronic Engineering
- BTEC National Diploma in Motor Vehicle Engineering
- foundation year for an Engineering Degree
- an Access to Engineering programme.

We have tried to provide a sympathetic approach to the fundamental mathematics required by engineers – it is then developed by application to real engineering situations, giving the underpinning background knowledge needed.

We hope you enjoy using this book and that it will really become a student's companion!

PART ONE

Use of number and algebra to solve engineering problems

The title of this part may seem a little confusing at first sight, but number and algebra are basic to all mathematical problems, regardless of whether you are an engineer or a bank manager.

Without the use of number we could not record or calculate. How would you express the cost of designing a motor car, or even the speed at which it is travelling without resorting to the use of number?

What is algebra and why is it needed? In algebra we use letters to express relationships, such as equations, from which number values are often obtained. You will already be familiar with expressions such as for a rectangle, Area = length × breadth or, using symbols, $A = l \times b$. This often simplifies engineering problems but, like everything, you have first to learn the rules. This book covers the key mathematical areas where number and algebra occur and their application in engineering situations.

The four chapters in this first part of the book build on each other. We start with number—units of measure, accuracy and calculation, number systems, then areas and volumes of the more common engineering components. Next comes algebra with the basic skills of manipulation using indices and logarithms as key tools. Finally, we have the application of algebra, using equations, to real practical problems and to finding numerical solutions.

1 Numbers in engineering problems

In this chapter we have five main sections, all of which are important if you are to be successful as an engineer.

Units of measure. Here we look at the different units of measure used in industry. First we cover the standard units, and then the way they are modified for use with very large or very small quantities. We must also be able to convert from units of one system to those of another, for example, from inches to millimetres.

Accuracy. How accurate is your work? It is important that we, as engineers, can ensure that the accuracy of our work matches that needed for the application. Does the error of one component affect the relative error of the final product?

Use of a calculator. A poor workman blames his tools, but a skilled workman looks after and understands the capability of his equipment. Here we try to ensure that you can first look at a computation and estimate an answer—a rough check. Then using your calculator will provide an accurate result—the rough check will usually show up any slip in the use of the machine which would otherwise be disastrous.

Number systems. Have we always used a ten digit number system? What other systems are there? An octal system may be used or the binary number system, which is favoured in the arithmetic of computers and electronic equipment. It is important that we can understand these systems if we are to make use of this technology.

Perimeters, areas and volumes. Most complicated problems can be solved using simple mathematical formulae. Here our knowledge of basic geometry will enable us to break down awkward areas and volumes into simple component shapes, so that we can proceed with calculations with confidence.

Units of measure

There are two main unit systems in engineering. They are:

a) The Système International d'Unités (the international system of units), which is usually abbreviated to SI and is similar to the metric system (using metres etc.)

b) The old British 'imperial' system (using feet and inches etc.).

SI units

Base units

We will look now at the six fundamental (or base) units, so called because they are fundamental to the whole system of measurement:

	Base unit	Abbreviation
Length	metre	m
Mass	kilogram	kg
Time	second	s
Electric current	ampere	A
Luminous intensity	candela	cd
Temperature	kelvin	K

Multiples and submultiples

Sometimes in engineering measurements are too big or too small to be measured using the base unit, so we use multiples and submultiples of 10 (10 being the number base of SI). These are known as prefixes and are shown in front of the base unit, and are given special names as follows:

Multiplication Factor		Prefix	Symbol
1 000 000 000 000	$= 10^{12}$	tera	T
1 000 000 000	$= 10^{9}$	giga	G
1 000 000	$= 10^{6}$	mega	M
1 000	$= 10^{3}$	kilo	k
100	$= 10^{2}$	hecto	h
10	$= 10^{1}$	deca	da
0.1	$= 10^{-1}$	deci	d
0.01	$= 10^{-2}$	centi	c
0.001	$= 10^{-3}$	milli	m
0.000 001	$= 10^{-6}$	micro	μ
0.000 000 001	$= 10^{-9}$	nano	n
0.000 000 000 001	$= 10^{-12}$	pico	p
0.000 000 000 000 001	$= 10^{-15}$	femto	f
0.000 000 000 000 000 001	$= 10^{-18}$	atto	a

The choice of which unit to use will often be yours, but there are some other guidelines, for example, distances between towns on a road map are in kilometres. In order to avoid manufacturers of components having to make too many variations, preferred units are used.

Preferred units

Preferred units are those which have the multiplication factors: $10^{12}, 10^{9}, 10^{6}, 10^{3}, 10^{-3}, 10^{-6}, 10^{-9}, 10^{-12}, 10^{-15}$ and 10^{-18} together with the base unit.

Thus, 5000 metres should be written as 5 kilometres, i.e. 5 km which is 5×10^{3} metres, and NOT as 50 hectometres, i.e. 50 hm which is 5×10^{2} metres.

Length (symbol *l*)

The SI base unit of length is the metre, which is suitable for work-shop or plan sites, but too large for plate thickness and too small for geographical distances, so the use of multiples and submultiples is needed. For

large distances, the kilometre	$1 \text{ km} = 10^{3} \text{ m}$
small lengths, the millimetre	$1 \text{ mm} = 10^{-3} \text{ m}$
tiny lengths, the micrometre (or micron)	$1 \, \mu\text{m} = 10^{-6} \text{ m}$

The above are a selection of preferred units—one non-preferred unit in common use for small distances is the centimetre where $1 \text{ cm} = 10 \text{ mm} = 10^{-2}$ m.

EXAMPLE 1.1

The shortest (great circle) distance between Cape Town and New York is 12 551 000 metres. Express this measurement in suitable preferred units.

Now $12 551 000 \text{ m} = 12 551 \times 10^{3} \text{ m} = 12 551$ kilometres (km)

$= 12.551 \times 10^{6} \text{ m} = 12.551$ megametres (Mm)

5

EXAMPLE 1.2

An extremely small aperture has width 0.000 000 82 metres. Express this measurement in suitable preferred units.

$$\text{Now } 0.000\,000\,82 \text{ m} = 0.000\,82 \times 10^{-3} \text{ m} = 0.000\,82 \text{ millimetres (mm)}$$
$$= 0.82 \times 10^{-6} \text{ m} \qquad = 0.82 \text{ microns } (\mu\text{m})$$
$$= 820 \times 10^{-9} \text{ m} \qquad = 820 \text{ nanometres (nm)}$$

Area (symbol A)

The basic unit of area is the square metre (m^2). Also for

large areas: the square kilometre $\qquad\qquad 1 \text{ km}^2 = 10^6 \text{ m}^2$
field areas:
   the square hectometre (or hectare) $\qquad 1 \text{ hm}^2 \text{ (or 1 ha)} = 10^4 \text{ m}^2$
lesser areas:
   the square decametre (or are) $\qquad\qquad 1 \text{ dam}^2 \text{ (or 1 a)} = 10^2 \text{ m}^2$
small areas: the square centimetre $\qquad\qquad 1 \text{ cm}^2 = 10^{-4} \text{ m}^2$
tiny areas: the square millimetre $\qquad\qquad 1 \text{ mm}^2 = 10^{-6} \text{ m}^2$

Although non-preferred, the units hectare, are, and square centimetre are in common use.

EXAMPLE 1.3

Find, in square metres, the area of a rectangular metal sheet measuring 1840 mm by 730 mm.

$$\text{Now} \quad 1840 \text{ mm} = 1.84 \times 10^3 \text{ mm} \quad \text{and} \quad 730 \text{ mm} = 0.73 \times 10^3 \text{ mm}$$

$$\begin{aligned}
\text{Thus} \quad \text{sheet area} &= (1.84 \times 10^3) \times (0.73 \times 10^3) \text{ mm}^2 \\
&= (1.84 \times 0.73) \times 10^{3+3} \text{ mm}^2 \\
&= 1.34 \times 10^6 \text{ mm}^2 \\
&= 1.34 \text{ m}^2 \quad \text{correct to 3 s.f.}
\end{aligned}$$

Volume (or capacity) (symbol V)

The basic unit of volume is the cubic metre (m^3)—this is suitable for most large volumes. For smaller volumes we use:

everyday measure,
   the cubic decimetre (or litre*) $\qquad\qquad 1 \text{ dm}^3 \text{ (or } 1\,\ell) = 10^{-3} \text{ m}^3$
small measure,
   the cubic centimetre (or millilitre*) $\qquad 1 \text{ cm}^3 \text{ (or 1 m}\ell) = 10^{-6} \text{ m}^3$
tiny volumes,
   the cubic millimetre $\qquad\qquad\qquad\qquad 1 \text{ mm}^3 = 10^{-9} \text{ m}^3$

*The litre (ℓ) has become the common unit for liquid measure, and to four-figure accuracy may be treated as $1\,\text{dm}^3$, but for any precise calculation the relationship between the litre and cubic decimetre is 1 litre $= 1.000\,028\,\text{dm}^3$.

EXAMPLE 1.4

A large tool-box measures 1330 mm by 650 mm by 550 mm. Find its volume in cubic metres.

$$\text{Now} \quad \text{box volume} = (1.33 \times 10^3) \times (0.65 \times 10^3) \times (0.55 \times 10^3) \text{ mm}^3$$
$$= (1.33 \times 0.65 \times 0.55) \times 10^{3+3+3} \text{ mm}^3$$
$$= 0.475 \times 10^9 \text{ mm}^3$$
$$= 0.475 \text{ m}^3 \text{ correct to 3 s.f.}$$

Mass (symbol m)

The SI base unit of mass is the kilogram, suitable for everyday use. For

large mass, the tonne (the metric ton) $1 \text{ t} = 10^3 \text{ kg}$
small mass, the gram $1 \text{ g} = 10^{-3} \text{ kg}$
tiny mass, the milligram $1 \text{ mg} = 10^{-3} \text{ g} = 10^{-6} \text{ kg}$
minute mass, the microgram $1 \text{ } \mu g = 10^{-6} \text{ g} = 10^{-9} \text{ kg}$

The tonne (metric ton) is very nearly equal to an imperial ton. The symbol, t, is used only for the tonne; the imperial ton should be written in full.

Exercise 1.1

Express more briefly each of the following with a suitable preferred unit.

1) 8000 m 2) 15 000 kg 3) 3800 km
4) 1800 000 kg 5) 0.007 m 6) 0.000 001 3 m
7) 0.028 kg 8) 0.000 36 km 9) 0.000 064 kg
10) 0.0036 A

Density (symbol ρ)

Density is defined as mass per unit volume. Since 'per' means 'divided by' we have

$$\text{Density} = \frac{\text{mass}}{\text{unit volume}} \quad \text{kilograms per cubic metre}$$

or in symbols $\rho = \dfrac{m}{V}$ kg/m^3 or $\dfrac{\text{kg}}{\text{m}^3}$ or   kg m^{-3}

Density of water   From the original metric definition

1 cm^3 of water has a mass of 1 g

so $1000 \text{ cm}^3 = 1\ell$ of water has a mass of $1000 \text{ g} = 1 \text{ kg}$

$1000 \quad \ell = 1 \text{ m}^3$ of water has a mass of $1000 \text{ kg} = 1 \text{ t}$

and

> A litre of water has a mass of one kilogram
> A cubic metre of water has a mass of one tonne

Density of common materials The densities of common materials may be found in engineering tables. A selection are given here:

Material	Density (kg/m^3)	Material	Density (kg/m^3)
Aluminium	2700	Mercury	13 600
Concrete	2200	Oak	700
Copper	8800	Oil (heavy)	900
Ice	900	Petrol	720
Iron (cast)	7200	Steel	7900
Lead	11 400	Water	1000

EXAMPLE 1.5

Find the mass of a concrete floor 5 m by 6 m by 50 mm thick, if the density of concrete is 2200 kg/m^3.

Now we know

$$m = \rho V$$

$$= 2200 \times \left(5 \times 6 \times \frac{50}{1000}\right) \frac{\text{kg}}{\text{m}^3} \times \text{m}^3$$

$$= 3300 \text{ kg}$$

Velocity (symbol v)

> Velocity is the rate of change of distance with respect to time.
> It has a derived unit metres per second (m/s or $\frac{\text{m}}{\text{s}}$ or m s^{-1}).
>
> Another common unit is kilometres per hour (km/h or $\frac{\text{km}}{\text{h}}$ or km h^{-1}).

EXAMPLE 1.6

Find the value of 20 m/s in km/h.

Our problem is to express $20\,\dfrac{\text{m}}{\text{s}}$ as $?\,\dfrac{\text{km}}{\text{h}}$

Now 1000 m = 1 km so $\left(\dfrac{1 \text{ km}}{1000 \text{ m}}\right) = 1$: also $\left(\dfrac{60 \times 60 \text{ s}}{1 \text{ h}}\right) = 1$

If we multiply anything by 1 it will remain unchanged in value, so:

$$20\,\frac{\text{m}}{\text{s}} = 20\,\frac{\text{m}}{\text{s}} \times \left(\frac{1 \text{ km}}{1000 \text{ m}}\right) \times \left(\frac{60 \times 60 \text{ s}}{1 \text{ h}}\right)$$

$$= \frac{20 \times 60 \times 60}{1000}\,\frac{\text{km}}{\text{h}}$$

$$\therefore \qquad 20\,\text{m/s} = 72 \text{ km/h}$$

Note how, by careful choice of the unity brackets, we arranged for the m units on the top line to be cancelled out and replaced by km units. Similarly, the s units on the bottom line to be replaced by h units. This choice is not luck or magic but something you will learn to do with practice.

Acceleration (symbol *a*)

Acceleration is the rate of change of velocity with respect to time. Its units are metres per second per second ($m/s/s$ or $\dfrac{m}{s^2}$ or ms^{-2})

EXAMPLE 1.7

A vehicle increases speed, at a uniform rate, from $30\,km/h$ to $60\,km/h$ in a time of 8 seconds. Find the vehicle's acceleration.

Now because the increase in velocity is uniform, the acceleration is constant.

$$\text{Thus} \qquad \text{constant acceleration} = \frac{\text{change in velocity}}{\text{time}}$$

$$= \frac{\text{final velocity} - \text{initial velocity}}{\text{time}}$$

or, in symbols,
$$a = \frac{v - u}{t}$$

Now we need to change the velocity units from km/h to m/s.

$$\text{Thus} \quad u = 30\,\frac{km}{h} = 30\,\frac{\cancel{km}}{\cancel{h}} \times \frac{1000\,m}{1\,\cancel{km}} \times \frac{1\,\cancel{h}}{3600\,s} = 8.33\,m/s$$

Also since *v* was given as twice *u*,

Then
$$v = 2 \times 8.33 = 16.66\,m/s$$

Hence
$$a = \frac{16.66 - 8.33}{8}\,\frac{m/s}{s}$$

$$= 1.04\,m/s/s \quad \text{or} \quad m/s^2 \quad \text{or} \quad m\,s^{-2}$$

Force (symbol *F*)

Newton's second law of motion, providing we have a coherent system of units, such as SI, may be simplified to:

Force = mass × acceleration

Now in SI the unit of force is the newton (N) derived from the statement:

1 N is the force which will accelerate 1 kg mass at a rate of $1\,m/s^2$

thus in symbols $\quad F\,(N) = m\,(kg) \times a\,(m/s^2)$

So, just looking at the units

$$1\,\text{N} = 1\,\text{kg} \times 1\,\text{m/s}^2$$

or
$$1\,\text{N} = 1\,\frac{\text{kg m}}{\text{s}^2}$$

So N units may be replaced by $\dfrac{\text{kg m}}{\text{s}^2}$ or kg m/s^2 or kg m/s^{-2} units and vice versa.

EXAMPLE 1.8

Find the force needed to give a mass of 600 kg an acceleration $5\,\text{m/s}^2$.

Now
$$F = m \times a$$
$$= 600\,\text{kg} \times 5\,\text{m/s}^2$$
$$= 600 \times 5\,\text{kg m/s}^2$$
$$= 3000\,\text{N}$$

Weight (symbol *W*)

Weight is simply the force of gravity acting on a body. On Earth it attracts the body towards the centre of the Earth; on the Moon it attracts the body towards the centre of the Moon, but to a lesser extent (gravity on the Moon is approximately one sixth of that on Earth).

So using   Force = mass × acceleration

then   Weight = mass × (acceleration of a
                                          freely falling body)

or, in symbols, W (N) $= m$ (kg) $\times g$ (m/s^2)

On earth the average value of g is $9.81\,\text{m/s}^2$ (often taken as $10\,\text{m/s}^2$ without too much error).

Thus, on earth,   1 kilogram weighs 9.81 newtons

SI calculations use mass in kilograms almost exclusively. Weight in newtons is only needed when, for example, calculating stresses in a weight carrying structure, or determining the lift required from an aircraft.

The general public, unaware of any difference between the two, will continue wrongly to refer to mass as weight, in phrases like 'my luggage weighs 20 kilograms'. So, probably, will the engineer, except when he or she has to design a luggage rack! Then he or she will recall that it is the mass which is 20 kilograms; its weight is 20×9.81 or 196 newtons.

EXAMPLE 1.9

What is the mass of a vehicle which weighs 9500 N?

From $\qquad$ Force $=$ mass $\times$ acceleration

then $\qquad W = m \times g$

$\therefore \qquad m = \dfrac{W}{g}$

$\qquad = \dfrac{9500}{9.81} = 968 \text{ kg}$

Pressure (symbol p)

Pressure (or stress) $= \dfrac{\text{force}}{\text{area on which it acts}}$

or in symbols $\qquad p = \dfrac{F}{A}$

The SI unit of pressure is the newton per square metre (N/m^2 or Nm^{-2}) and is sometimes called the pascal (Pa).

A more useful unit is the bar, since atmospheric pressure is almost exactly equal to 1 bar (or 1.013 25 bar exactly, international standard).

Thus $\qquad$ 1 bar $=$ 1 atmosphere $=$ 100 000 $\quad$ or $\quad 10^5 \text{ N/m}^2$

EXAMPLE 1.10

Find the force on a rectangular sheet of plastic 200 mm by 300 mm if a pressure of 0.9 kN/m^2 is applied.

Now $\qquad$ pressure $= \dfrac{\text{force}}{\text{area}}$

Thus $\qquad$ force $=$ pressure $\times$ area

Converting to basic SI units: $0.9 \text{ kN/m}^2 = 0.9 \times 1000 \text{ N/m}^2$,

so $\qquad$ force $= (0.9 \times 1000) \times \left(\dfrac{200}{1000} \times \dfrac{300}{1000} \right)$

$\qquad = 54 \text{ N}$

Energy

Energy is measured in joules, and 1 joule is the work done when a force of 1 newton acts through a distance of 1 metre. Assuming no losses, this amount of work would be capable of generating 1 watt of electricity for 1 second. Thus:

$\qquad$ 1 joule $=$ 1 watt second $=$ 1 newton metre

or $\qquad$ 1 J $=$ 1 Ws $=$ 1 Nm

The joule is also the unit of heat and the unit of electrical energy.

Power (mechanical)

Power is the rate of doing work   or $\dfrac{\text{work done}}{\text{time}}$

In SI the unit of power is the watt (W). A mechanical power of 1 watt is a rate of working of 1 joule per second. Thus:

$$1\,\text{W} = 1\,\frac{\text{J}}{\text{s}} = 1\,\frac{\text{Nm}}{\text{s}}$$

EXAMPLE 1.11

In a belt drive the belt velocity is 740 metres per minute and the power transmitted is 64 kW. Find the difference in tension between the tight and slack sides of the belt.

Now   power $= \dfrac{\text{work done}}{\text{time}} = \dfrac{\text{force} \times \text{distance}}{\text{time}} = \text{force} \times \dfrac{\text{distance}}{\text{time}}$

$\qquad\qquad\qquad = \text{force} \times \text{velocity}$

From which $\qquad\qquad\qquad$ force $= \dfrac{\text{power}}{\text{velocity}}$

$\qquad\qquad\qquad\qquad\qquad = \dfrac{64 \times 1000}{740/60}\,\dfrac{\text{W}}{\text{m/s}}$

so $\qquad$ Difference in tensions $= 5190\,\text{N}$ since all data used was converted into basic SI units.

Power (electrical)

In SI the unit of electrical power and the unit of mechanical power are one and the same: the watt. Thus, if there were no losses in the conversion from one form to the other, the work input to a dynamo at the rate of 1 newton metre per second would generate 1 watt of electricity. Similarly, an input to an electric motor of 1 watt of electricity would cause it to do work at the rate of 1 Nm/s or 1 J/s.

In direct current circuits (or alternating current circuits where the voltage and current are in phase):

$\qquad$ power (watts) $=$ current (amperes) $\times$ voltage (volts)

EXAMPLE 1.12

An electric motor has an efficiency of 90% and gives an output of 12 kW when connected to a 240 V supply. Calculate

a) the current taken, and

b) the energy over a three hour run.

a) Now

$$\text{efficiency} = \frac{\text{output}}{\text{input}} \times 100\%$$

so

$$\text{input power} = \frac{\text{output}}{\text{efficiency}} \times 100\%$$

$$= \frac{12 \times 1000}{90} \times 100\%$$

$$= 13\ 333 \text{ W}$$

Also

$$\text{power} = \text{current} \times \text{voltage}$$

so

$$\text{input current} = \frac{\text{input power}}{\text{voltage}}$$

$$= \frac{13\ 333}{240}$$

$$= 55.6 \text{ A}$$

b) Now

$$\text{energy (joules)} = \text{power (watts)} \times \text{time (seconds)}$$

so   energy input for 3 hours

$$= 13\ 333 \times (3 \times 60 \times 60) \text{ J}$$

$$= 1.44 \times 10^8 \text{ J}$$

$$= 144 \text{ MJ}$$

Electric charge (symbol Q)

Now

$$\text{charge} = \text{current} \times \text{time}$$

or in symbols

$$Q = I \times t$$

The unit of charge is the coulomb (C) and is defined as the quantity of electricity which flows past a certain point in an electric circuit when a current of 1 ampere is maintained for 1 second.

Electrical resistance (symbol R)

The resistance in an electric circuit is its opposition to the flow of electric current.

The unit of resistance is the ohm (Ω), which is the resistance between two points of a conductor when a potential difference of 1 volt applied between these points produces a current of 1 ampere.

Ohm's law gives

$$\text{current (A)} = \frac{\text{voltage (V)}}{\text{resistance } (\Omega)}$$

or in symbols

$$I = \frac{V}{R}$$

so

$$V = IR \quad \text{or} \quad R = \frac{V}{I}$$

EXAMPLE 1.13

a) What is the voltage difference across a $8\,k\Omega$ resistor when a current of $80\,\mu A$ is flowing?

b) What charge is transferred if the current flows for 14 hours?

a) Now from Ohm's law $V = IR$

so, in basic SI units, $8\,k\Omega = 8 \times 10^3\,\Omega$ and $80\,\mu A = 80 \times 10^{-6}\,A$

giving $$V = (80 \times 10^{-6}) \times (8 \times 10^3)$$
$$= 640 \times 10^{-3} = 0.64\,V$$

b) Now we know $Q = It$

so, working in basic SI units of amperes and seconds,

then, $$Q = (80 \times 10^{-6}) \times (14 \times 60 \times 60) = 4.032\,C$$

Exercise 1.2

1) Electric voltage V (units V, volts), current I (units A, amperes) and resistance R (units Ω, ohms) are connected by the expression $V = IR$. Find the voltage if $I = 15\,\mu A$ and $R = 30\,k\Omega$.

2) The equation $F = ma$ relates force F (units N, newtons), mass m (units kg, kilograms) and acceleration a (units ms^{-2}, metres per second per second).
   a) If a 1 tonne vehicle accelerates at $2\,ms^{-2}$ find the force required.
   b) A motor-car weighing $8000\,N$ slows down due to an average braking force of $3.2\,kN$. Find the deceleration of the vehicle.

3) The work done by a force F (units N, newtons) moving through a distance s (units m, metres) is given by Fs (units J, joules). Find the work done in kJ if a force of $300\,N$ moves through a distance of $0.1\,km$.

4) Power, in mechanics, is the rate of doing work and is given by $\dfrac{\text{work done}}{\text{time}}$ or in symbols $\dfrac{Fs}{t}$ or $F \times \dfrac{s}{t}$ which is Fv, where v is the velocity in units ms^{-1}. Using standard basic SI units, power is measured in units W, watts.

   Find the power in kW when a driving force of $5\,kN$ (being the difference in tensions between the tight and slack sides) of a belt drive is moving with a belt speed of $12\,ms^{-1}$.

5) Electrical power is also measured in W, watts and can be found by multiplying current I (units A, amperes) by voltage V (units V, volts), or using symbols, IV. Find the power of an electric fire which takes a current of $8\,A$ from a $250\,V$ mains supply.

6) Kinetic energy is the energy due to the motion of a mass and is given by $\frac{1}{2}mv^2$ (units J, joules) where mass m kg has velocity v ms^{-1}. What is the kinetic energy in MJ units possessed by a motor-car which has a mass of 0.8 tonne, when travelling at a speed of $60\,km\,h^{-1}$?

7) Calculate the work done, in kJ, in raising a 500 kg rolled steel joist by a crane from ground level to the top floor of a building, 30 m above. Also what is the power need if this takes place in 30 seconds?

Imperial system of units

This system still exists for a number of reasons:

1) It has been in place throughout the development of engineering in Great Britain.

2) People have resisted, and found difficulty in accepting, a new system.

3) The cost of replacing equipment and machinery would be large.

4) There is a need to produce imperial components for existing plant.

Conversions to SI units

We must be able to convert units from one system to the other, and so conversion factors are listed below with the more common imperial units.

Length	Conversion factors (correct to 3 s.f.)
12 inches (in) = 1 foot (ft)	1 in = 25.4 mm
3 feet (ft) = 1 yard (yd)	1 ft = 0.305 m
1760 yards (yd) = 1 mile	1 mile = 1.61 km
Area	
1 acre = 4840 yd^2	1 acre = 0.405 ha
Volume (capacity)	
8 pints = 1 gallon (gal)	1 gal = 4.55 ℓ
Mass	
16 ounces (oz) = 1 pound (lb)	1 lb = 0.454 kg
112 pounds (lb) = 1 hundredweight (cwt)	1 ton = 1020 kg
20 cwt = 1 ton	or 1 ton $\approx$ 1000 kg
	or 1 tonne (t)

EXAMPLE 1.14

Express 5.24 yards in metres.

Since $3\,\text{ft} = 1\,\text{yd}$,   and $0.305\,\text{m} = 1\,\text{ft}$

then

$$5.24\ \text{yd} = 5.24\ \cancel{\text{yd}} \times \left(\frac{3\ \cancel{\text{ft}}}{1\ \cancel{\text{yd}}}\right) \times \left(\frac{0.305\ \text{m}}{1\ \cancel{\text{ft}}}\right)$$

$$= 4.79\ \text{m correct to 3 s.f.}$$

EXAMPLE 1.15
Change 72 kilograms into hundredweight.

Since   1 lb = 0.454 kg,   and   1 cwt = 112 lb

then
$$72 \text{ kg} = 72 \cancel{\text{kg}} \times \left(\frac{1 \cancel{\text{lb}}}{0.454 \cancel{\text{kg}}} \right) \times \left(\frac{1 \text{ cwt}}{112 \cancel{\text{lb}}} \right)$$
$$= 1.42 \text{ cwt correct to 3 s.f.}$$

EXAMPLE 1.16
What is a pint of beer measured in litres? Also how many pints would I get if I ordered a litre of beer?

Since   1 gal = 8 pt,   and   4.55 ℓ = 1 gal

$$\text{then} \quad 1 \text{ pt} = 1 \cancel{\text{pt}} \times \left(\frac{1 \cancel{\text{gal}}}{8 \cancel{\text{pt}}} \right) \times \left(\frac{4.55 \, \ell}{1 \cancel{\text{gal}}} \right)$$

giving   1 pt = 0.57 litres correct to 2 s.f.

Now if I said that $1 \, \ell = \frac{1}{0.57}$ pt from the result above, I would arrive at 1 ℓ = 1.75 pt but I cannot say that this is correct to 3 s.f., since the 0.57 was rounded correct to only 2 s.f. Thus I must either content myself by saying that 1 litre = 1.8 pints or start afresh with conversion factors correct to 3 s.f.

Exercise 1.3

Give the answers to 3 s.f. unless there is a good reason for doing otherwise.

1) The dimensions of a factory workshop are being converted from imperial to SI units. One of these measurements is 880 yards and its equivalent in metre units is needed.

2) Slip gauges are required for a measurement of 2.16 inches. As only metric ones are available, what is this dimension in millimetres?

3) A component has a surface area of 4.2 square feet. Since painting costs are quoted per square metre, how many square metres are there?

4) A spherical gas container has a volume of 8 ft^3. A customer wishes to know its capacity in m^3.

5) A milk marketing company wishes to know how many pints of milk can be transported in a container we manufacture which has a nominal capacity of 25 litres.

6) A motor van has a mass of 25 cwt. What figures should be used for the vehicle mass on a registration form stating kilogram units?

7) A development department wish to know how many ml of an extremely high grade and expensive oil are required to fill a reservoir for a bearing of a heavy machine – its capacity, etched on the side of the reservoir, is 3 in^3.

8) A drawing dimension is given as 0.906 ± 0.0012 inches. Convert this to mm.

9) Three important dimensions of a 6 BA thread are: diameter 2.8 mm, pitch 0.53 mm, and depth of thread 0.0125 mm. Convert these to inches.

10) A tank is to hold 150 gallons.
    a) Find its volume in ft³.
    b) How many litres does the tank hold?
    c) What is the volume in m³?

11) How much must be ground off a plug gauge 1.625 inches in diameter so that it can be used for checking a hole of 41 mm diameter? Give your answer to the nearest 0.001 in.

12) Express a speed of 60 mile/hour as
    a) ft/s   b) m/s   c) km/h.

13) A vehicle travels 30 miles to the gallon of petrol. How many kilometres will it travel on 4 litres of petrol?

Approximation and accuracy

In engineering the majority of numbers are not exact or discrete. An example of an exact number would be the thirteen persons employed by the local motor main agents. This number is exact or discrete, and therefore cannot be an approximation. It is often necessary to approximate any non-exact answers we calculate, to the level of precision required. This will depend on the particular application: for instance, if we are ordering raw materials it is likely that measurements to the nearest millimetre would be good enough. However, when manufacturing a component for a motor vehicle engine we may well be thinking in terms of thousandths of a millimetre.

Accuracy of numbers

There are two principal methods of expressing the accuracy of a number. These are using either decimal places or significant figures, together with rounded numbers.

Decimal places (abbreviated to 'd.p.')

These refer to the number of figures which follow (i.e. after or to the right of) the decimal point.

Thus: 35.1 has one d.p. and 2.402 has three d.p.

Significant figures (abbreviated to 's.f.')

These are the number of figures, counted from the left to the right, starting with the first *non-zero* figure, unless stated otherwise.

Thus: 2700, 35.0, 0.89 and 0.0082 each have two s.f.
          354, 7.21 and 0.000 234 each have three s.f.
          5 782 100 and 0.537 91 each have five s.f.

Note that zero figures at the right hand end are not included in the count, as in 2700, 35.0 and 5 782 100 mentioned above.

Rounded numbers

These are obtained by the process of 'rounding' or 'rounding-off' and enable a degree of accuracy to be stated.

Thus: rounding 31.63 gives 31.6 correct to three s.f.
and   rounding 31.68 gives 31.7 correct to three s.f.

> **Rule for rounding**   Working from right to left figures are discarded in turn:
>
> If the discarded figure is less than 5 the preceding figure is *not* altered—in the first example given above the 3 is discarded and the 6 is unaltered. If the discarded figure is greater than (or equal to) 5 then the preceding figure is increased by one—in the second example given above the 8 is discarded and the preceding figure 6 is increased to 7.
>
> Thus 0.472 becomes 0.47   correct to 2 s.f.
> or 0.5   correct to 1 s.f.
>
> Also 24.0926 becomes 24.093   correct to 3 d.p.
> or 24.09   correct to 2 d.p.
> or 24.1   correct to 1 d.p.

An exception to the rule is shown below:

Now   0.008 246   may be stated as
           0.008 25   correct to 3 s.f.; alternatively to 5 d.p.
or   0.0082   correct to 2 s.f.; alternatively to 4 d.p.
or   0.008   correct to 1 s.f.; alternatively to 3 d.p.

Care must be taken to ensure that rounding is carried out directly from the original number—thus in the above case 0.008 246 rounds to 0.0082 correct to 2 s.f.

It would be wrong to round successively to 3 s.f. and then to 2 s.f.: the second rounding of 0.008 25 would give an incorrect 0.0083 !

Note: when rounding whole number amounts, 'discarded' figures preceding the decimal point must be replaced by zeros.

Thus: 1479 becomes   1480 (*not* 148)   correct to 3 s.f.
or   1500 (*not* 15)   correct to 2 s.f.
or   1000 (*not* 1)   correct to 1 s.f.

Consider an attendance of 54 276 at a league soccer match. This is as precise as could have been obtained, no doubt from the turnstiles at entry.

Now 54 276 may be stated as   54 280   correct to 4 s.f.
or   54 300   correct to 3 s.f.
or   54 000   correct to 2 s.f.
or even   50 000   correct to 1 s.f.

As far as press reports are concerned 54 300 may well be considered to be good enough. So when the papers state an attendance of 54 300 a 3 s.f. accuracy is implied (although it would not be mentioned!).

Exercise 1.4

Write down the following numbers correct to the number of significant figures stated:

1)	24.865 82	a)	to 6	b)	to 4	c)	to 2
2)	0.008 3571	a)	to 4	b)	to 3	c)	to 2
3)	4.978 48	a)	to 5	b)	to 3	c)	to 1
4)	21.987 to 2						
5)	35.603 to 4						
6)	28.387 617	a)	to 5	b)	to 2		
7)	4.149 76	a)	to 5	b)	to 4	c)	to 3
8)	9.2048 to 3						

Write down the following numbers correct to the number of decimal places stated:

9)	2.138 87	a)	to 4	b)	to 3	c)	to 2
10)	25.165	a)	to 2	b)	to 1		
11)	0.003 988	a)	to 5	b)	to 4	c)	to 3
12)	7.2039	a)	to 3	b)	to 2	c)	to 1
13)	0.7259	a)	to 3	b)	to 2		

Errors in numbers

In technology it is important that we know, and are able to state clearly, the exact accuracy of a value. A convenient way is to state the amount of allowable error in a measurement.

A typical, non-exact, measurement is a person's height of 117.8 cm (an approximation to 4 s.f., or to the nearest $\frac{1}{10}$ cm).

Now consider an electrical resistance measured as 52 ohm, to the nearest ohm, and considered correct to 2 s.f. The number 52 could have been obtained by rounding any number between the lowest value of 51.5 and up to the highest value of 52.5. These extremes are $52 - 0.5$ and $52 + 0.5$ and the value of the resistance would be stated as 52 ± 0.5 ohm. This shows, at a glance, the maximum possible error is 0.5 greater, or 0.5 smaller, than 52. This does not mean that the error is bound to be as great at 0.5 ohm—it may be considerably less, but it certainly cannot be any more.

Similarly 7.49 gram is considered accurate to 3 s.f., or to $\frac{1}{100}$ gram: this would come from rounding a number between the extremes of 7.485 and 7.495 and would be given as 7.49 ± 0.005 gram.

Also 0.1370 mV given accurate to the nearest $\frac{1}{10\,000}$ mV is also considered accurate to 4 s.f. You should note how the extra zero is included to indicate accuracy greater than would be implied by 0.137 mV which is only correct to $\frac{1}{1000}$ mV and only accurate to 3 s.f. Now 0.1370 mV would come from rounding between the extremes of 0.136 95 mV and 0.137 05 mV and would be stated as $0.1370 \pm 0.000\,05$ mV.

Absolute and relative accuracy

Suppose we are told that a spring balance has an error of $\frac{1}{2}$ kg. This is an actual or true error value and is known as the absolute error.

How important is this? Well, this will depend on the mass being measured. It is unlikely that anyone would be bothered when the balance was used for 50 kg of potatoes, but when used for 2 kg of tomatoes such an error would not be acceptable.

In each case we are comparing the absolute error value with another measured value. We are, in fact, considering the error relative to the mass measured. You will now see the importance of a relative value.

Absolute accuracy

Absolute accuracy refers to the maximum absolute error of a number. Thus for the 52 ohm mentioned earlier, which if written in full is 52 ± 0.5 ohm, the maximum absolute error is 0.5 ohm. Thus the absolute accuracy of 52 ohm is 0.5 ohm.

Relative accuracy

The word relative means that the absolute error of a number must be compared with the number itself. This is usually given as a percentage using the expression:

$$\text{relative error} = \frac{\text{absolute error}}{\text{number}} \times 100\%$$

Relative accuracy is the phrase we use for the maximum relative error of a number.

So referring to the 52 ohm and the 0.5 ohm maximum absolute error

$$\text{maximum relative error} = \frac{0.5}{52} \times 100$$
$$= 0.96\% \text{ correct to 3 d.p.}$$

Thus, the relative accuracy of 52 ohm to the nearest ohm is 0.96% correct to 2 d.p.

Accuracy in addition and subtraction

A contractor has five vehicles. If he sells two he is left with three. So in this case of $5 - 2 = 3$ the answer 3 is exact, and there is no error, since the numbers 5 and 2 are discrete.

However, let us consider the result of adding electrical resistances of 52 ± 0.5 ohm and 36 ± 0.5 ohm.

We may state this problem as $(52 \pm 0.5) + (36 \pm 0.5)$.

$$
\begin{aligned}
\text{Now the greatest answer} &= \text{the greatest value of } 52 \pm 0.5 \\
&\quad + \text{ the greatest value of } 36 \pm 0.5 \\
&= (52 + 0.5) + (36 + 0.5) \\
&= (52 + 36) + (0.5 + 0.5) \\
&= (88 + 1.0) \text{ or } 89
\end{aligned}
$$

Similarly the smallest answer is $(88 - 1.0)$ or 87.

So the final result of adding the resistances lies between 87 and 89 ohm, and would be given as 88 ± 1.0 ohm, which has a maximum absolute error of 1.0 ohm.

> In general when adding and subtracting numbers the maximum absolute error of the result may be found by adding the maximum absolute errors of the original numbers.

$$
\begin{aligned}
\text{Thus } 623 + 56.3 \text{ implies} &\quad (623 \pm 0.5) + (56.3 \pm 0.05) \\
\text{giving a result} &\quad (623 + 56.3) \pm (0.5 + 0.05) \\
\text{or} &\quad 679.3 \pm 0.55
\end{aligned}
$$

$$
\begin{aligned}
\text{Now } 27.24 - 9.3 \text{ implies} &\quad (27.24 \pm 0.005) - (9.3 \pm 0.05) \\
\text{giving a result} &\quad (27.24 - 9.3) \pm (0.005 + 0.05) \\
\text{or} &\quad 17.94 \pm 0.055
\end{aligned}
$$

Note that even when the original numbers are being *subtracted* the maximum absolute error (0.055 in the above example) is still obtained by *adding* the given individual maximum absolute errors.

Accuracy in multiplication and division

Consider finding the area of a rectangle with sides measured as 67 mm and 62 mm. Each of the numbers will be considered accurate to two significant figures and so the problem may be stated as:

$$
\begin{aligned}
\text{Area} &= (67 \pm 0.5) \times (62 \pm 0.5) \\
\text{now} \quad \text{the greatest area} &= (\text{the greatest value of } 67 \pm 0.5) \\
&\quad \times (\text{the greatest value of } 62 \pm 0.5) \\
&= 67.5 \times 62.5 \\
&= 4218.75 \\
\text{and} \quad \text{the smallest area} &= (\text{the smallest value of } 67 \pm 0.5) \\
&\quad \times (\text{the smallest value of } 62 \pm 0.5) \\
&= 66.5 \times 61.5 \\
&= 4089.75
\end{aligned}
$$

If we examine these two extremes we see that only the four thousand figure is guaranteed in the value of the area.

Now it is generally accepted that:

> When multiplying and dividing numbers the answer should *not* be given to an accuracy greater than the least accurate of the given numbers.

If we simply calculate 67×62 we get 4154.

Here both the given lengths are to the same accuracy, so in this particular case either may be taken to be the least accurate, namely to two significant figures. Thus, after rounding, we have the area as 4200 mm², which would be an acceptable result.

You will possibly grumble at this answer and point out that even this is not strictly correct—and you would be right! Anyway, you now realise how wrong it is to give results to an accuracy which cannot be justified by the given data: we are all guilty of this from time to time so beware!

Consider also $\dfrac{5.73 \times 21}{0.6243}$ the result of which is 192.743 87 from a calculator. The least accurate of the three given numbers is 21 which has a two significant figure accuracy. Hence the answer must not be stated any more accurately than this: namely 190 obtained by rounding 192.743 87 correct to two significant figures.

Implied accuracy

Suppose we decided to check graphically, by counting the squares, the area of the rectangle in the preceding section. Sides of 67 mm and 62 mm would be measured out, possibly with a rule, on squared paper. It is likely that we should try for, and achieve, a measuring accuracy to one tenth of a millimetre—thus the area we would be checking would be 67.0 mm by 62.0 mm. Now, since the calculated result of 67×62 is 4154, the area being checked would be 4150 mm² after rounding to three significant figures which is the accuracy of 67.0 mm and 62.0 mm.

With this in mind, unless there is a very good reason otherwise, a three significant figure accuracy is generally accepted on this type of data. We are really covering up for our inability to state the original data to its correct accuracy e.g. 67 mm which we should have given as 67.0 mm.

Consider, also, the problem of calculating the angles of a triangle with the sides given as 6, 8 and 9 m respectively. An accuracy of one significant figure (to which these dimensions are given) would only allow us to give the calculated angles such as 10°, 20°, 30°, ... etc. Again, we would assume the dimensions should have been given as 6.00, 8.00 and 9.00 m, and thus give the results to a three significant figure accuracy.

Truncation or cutting off

Some calculators 'truncate' or 'cut-off' figures in their displays after computation. For instance, in an eight figure display, the result of 5 divided by 3 would be shown as 1.666 666 66 (most modern machines would round the result to show 1.666 666 67). If truncation does occur each time successive computations are performed then an accumulating error may be introduced.

Exercise 1.5

In the questions below relative accuracies should be given to 3 s.f.

What are the greatest and least values, and also the relative accuracy, of

1) 64 ± 0.5

2) 2469 ± 5

3) 3.07 ± 0.005

4) 0.6 ± 0.05

What are the greatest and least values, and the absolute accuracy of the results, of

5) $(26 \pm 0.5) + (3.4 \pm 0.05)$

6) $(0.56 \pm 0.005) + (0.7 \pm 0.05)$

7) $(5.6 \pm 0.05) - (2.9 \pm 0.05)$

8) $(0.78 \pm 0.005) - (0.034 \pm 0.0005)$

9) A measurement has been taken of 39.07 km to the nearest $\frac{1}{100}$ km. State this in conventional form together with the absolute and relative accuracies.

10) A rectangle has sides of length 0.372 m and 1.238 m both measured to the nearest millimetre. State both these measurements in conventional form. Find also the maximum and minimum values of the perimeter of the rectangle, together with the absolute and relative accuracies of its nominal value.

11) Find the absolute and relative accuracies of the result of $(2.3 \pm 0.05) - (0.76 \pm 0.003) + (64 \pm 0.5)$

12) A current has been measured correct to $\frac{1}{100}$ A, but has been listed incorrectly as 7 A. How should it have been listed, and what is the maximum absolute error and the relative accuracy?

13) Careless recording gave the lengths of the sides of the triangle as 3 mm, 4 mm and 5 mm when in actual fact they had been measured to the nearest $\frac{1}{100}$ mm. What are the greatest and least values of the perimeter of the triangle? Also what are the absolute and relative accuracies of the perimeter value?

Use of a calculator

The calculator is an important tool for engineers and so it is important that we are familiar with its use. It enables us to complete numerical calculations quickly, accurately and reliably. However, we all make mistakes and we should therefore be able to recognise a blatantly wrong answer. How? By getting into the habit of doing a rough check first so you will have some idea of the result.

Rough checks

When using a calculator it is essential for you to do a rough check in order to obtain an approximate result. Any error, however small it may seem, in carrying out a sequence of operations will result in a wrong answer.

Suppose, for instance, that you had £1000 in the bank and then withdrew £97.82. The bank staff then used a calculator to find how much money you had left in your account—they calculated that £1000 less £978.2 left only £21.80 credited to you. You would be extremely annoyed and probably point out to them that a rough check of £1000 less £100 would leave £900, and that if this had been done much embarrassment would have been avoided.

The 'small' mistake was to get the decimal point in the wrong place when recording the money withdrawn, which is typical of errors we all make from time to time. You should get in the habit of doing a rough check on any calculation *before* using your machine. The advantage of a rough check answer before the actual calculation avoids the possibility of forgetting it in the excitement of obtaining a machine result. Also your rough check will not be influenced by the result obtained on your calculator.

Keyboard layout and operation

There are many different calculators on the market but they generally have similar functions. This section will look at the common features which apply to most scientific calculators. However, each calculator is supplied with an instruction booklet and you should work through this carrying out any worked examples which are given. Some operations may differ slightly from those given in this book. Part of a typical keyboard layout is shown in Fig. 1.1.

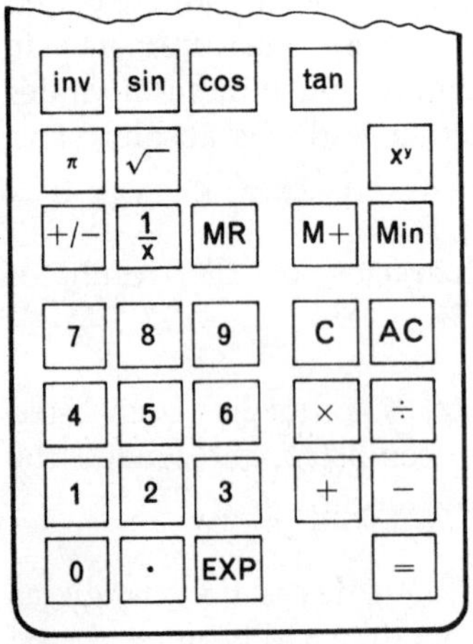

Fig. 1.1

The following are the figure keys:

$\boxed{0}\boxed{1}\boxed{2}\boxed{9}\boxed{4}\boxed{5}\boxed{6}\boxed{7}\boxed{8}\boxed{9}$

The other keys are summarised as follows:

$\boxed{\cdot}$ decimal point key.

$\boxed{\text{EXP}}$ exponent key – allows entry of numbers in standard form (e.g. 1.46×10^3)

$\boxed{\text{C}}$ clear key – enables an incorrect entry to be erased (either a figure or an operation) if pressed immediately afterwards.

$\boxed{\text{AC}}$ all clear key – clears machine of all numbers – except the memory.

$\boxed{\times}\boxed{\div}\boxed{+}\boxed{-}\boxed{=}$ arithmetical operation keys.

$\boxed{\text{Min}}$ memory in key – enters a number on display into the memory, erasing any number previously in the memory.

$\boxed{\text{M+}}$ add to memory key – enables a number to be added to previous content of memory.

$\boxed{\text{MR}}$ memory recall key – enables content of memory to be shown on display.

$\boxed{+/-}$ change sign key – (e.g. $+2$ to -2, or -1.5 to $+1.5$).

$\boxed{\frac{1}{x}}$ reciprocal key (e.g. $\frac{1}{2}$ to 2, or 4 to 0.25).

$\boxed{\pi}$ 'pi' key – gives the numerical value of π.

$\boxed{\sqrt{\ }}$ square root key.

$\boxed{x^y}$ power key – gives the value of x to the index y.

$\boxed{\text{sin}}\boxed{\text{cos}}\boxed{\text{tan}}$ trigonometric function keys – gives the sine, cosine or tangent of an angle.

$\boxed{\text{inv}}$ inverse trigonometric function key – gives the angle when given the value of a trigonometrical ratio, e.g. inv sin θ (also called arc sin θ or $\sin^{-1} \theta$).

Worked examples

After first switching on the calculator, or commencing a fresh problem, you should press the $\boxed{\text{AC}}$ key. This ensures that all figures entered previously have been erased and will not interfere with new data to be entered.

The memory is not cleared but this is done automatically when a new number is entered in the memory using the $\boxed{\text{Min}}$ key.

Numbers in standard form

If we have a very large or very small number it can be difficult to deal with for a number of reasons. First, if it has more than eight digits (often mostly zeros) it cannot be entered directly into a calculator. Then, the greater the number of digits, the more likely it is that errors will be made in manipulating the figures. So we show numbers like this in standard form. This means they are all displayed in the same way: a single digit 1 to 9 followed by the decimal places, which are then multiplied by a power of 10.

Let us consider $\quad 589\,000\,000\,000 = 5.89 \times 100\,000\,000\,000$

$$= 5.89 \times 10^{11} \text{ in standard form}$$

This tells us that the original number is 5.89 with the decimal point moved eleven places to the right.

Another example $\quad 0.000\,001\,2763 = \dfrac{1.2763}{1\,000\,000} = \dfrac{1.2763}{10^6}$

$$= 1.2763 \times 10^{-6} \text{ in standard form}$$

Here the index is negative, and so we must move the decimal point six places to the left to make 1.2763 smaller.

EXAMPLE 1.17
Find the value of $\quad 6\,857\,000 \times 119\,000 \times 85.3$

For the rough check numbers which contain as many figures as these are better considered in standard form:

$$(6.857 \times 10^6) \times (1.19 \times 10^5) \times (8.53 \times 10)$$

or approximately:

$$(7 \times 10^6) \times (1 \times 10^5) \times (10 \times 10) = 7 \times 1 \times 10 \times 10^{6+5+1}$$
$$= 70 \times 10^{12}$$
$$= 7 \times 10^{13}$$

The sequence of operation is:

$$\boxed{\text{AC}}\,\boxed{6}\,\boxed{8}\,\boxed{5}\,\boxed{7}\,\boxed{0}\,\boxed{0}\,\boxed{0}\,\boxed{\times}\,\boxed{1}\,\boxed{1}\,\boxed{9}\,\boxed{0}\,\boxed{0}\,\boxed{0}\,\boxed{\times}\,\boxed{8}\,\boxed{5}\,\boxed{\cdot}\,\boxed{9}\,\boxed{=}$$

The display will show $\boxed{6.96033\ 13}$ which represents $6.960\,33 \times 10^{13}$. The least number of significant figures in the given numbers is three. Thus the answer is 6.96×10^{13}.

An alternative method is to enter the numbers in standard form using the $\boxed{\text{EXP}}$ (exponent) key. The sequence would then be:

$$\boxed{\text{AC}}\,\boxed{6}\,\boxed{\cdot}\,\boxed{8}\,\boxed{5}\,\boxed{7}\,\boxed{\text{EXP}}\,\boxed{6}\,\boxed{\times}\,\boxed{1}\,\boxed{\cdot}\,\boxed{1}\,\boxed{9}\,\boxed{\text{EXP}}\,\boxed{5}$$
$$\boxed{\times}\,\boxed{8}\,\boxed{\cdot}\,\boxed{5}\,\boxed{9}\,\boxed{\text{EXP}}\,\boxed{1}\,\boxed{=}$$

giving the same result.

The sequence used in a problem such as this would be personal choice, but if the problem includes numbers with powers of 10 the latter sequence is better.

EXAMPLE 1.18

$$\text{Find the value of}: \quad 9.7 + \frac{55.15}{29.6 - 8.64}$$

$$\text{The rough check gives}: \quad 10 + \frac{60}{30 - 9} \approx 10 + \frac{60}{20} = 13$$

It is possible to work this problem out on the calculator by rearranging and using the memory. However no rearrangement is necessary if we make use of the $\boxed{\frac{1}{x}}$ (the reciprocal) key.

This key enables us to find the reciprocal of a number—for example, the reciprocal of 2 is $\frac{1}{2}$ or 0.5.

$$\text{Let us consider}\quad 9.7 + \cfrac{1}{\left(\cfrac{29.6 - 8.64}{55.15}\right)}$$

This may be written as:

$$9.7 + 1 \div \left(\frac{29.6 - 8.64}{55.15}\right) = 9.7 + 1 \times \left(\frac{55.15}{29.6 - 8.64}\right)$$

If we make use of this knowledge then the sequence of operations is:

The display gives $\boxed{12.331\ 202}$

It is always difficult to assess the accuracy of the answer to a calculation which involves addition and subtraction. Although the 9.7 has only two significant figures, the addition of the other portion of the calculation will increase the figures before the decimal point to two.

Thus the answer may be given as 12.3.

In most engineering problems you will not often be wrong if you give answers to three significant figures—this is consistent with the accuracy of much of the data (such as ultimate tensile strengths of materials). There are exceptions, of course, such as certain machine stop problems which may require a much greater degree of accuracy.

Exercise 1.6

Evaluate the following, take care to give the answers to an accuracy determined by the given data.

1) $45.6 + 3.5 - 21.4 - 14.6$

2) $-23.94 - 6.93 + 1.92 + 17.60$

3) $\dfrac{40.72 \times 3.86}{5.73}$

4) $\dfrac{4.86 \times 0.008\ 34 \times 0.640}{0.860 \times 0.934 \times 21.7}$

5) $\dfrac{57.3 + 64.29 + 3.17}{64.2}$

6) $\dfrac{32.2}{6.45 + 7.29 - 21.3}$

7) $\dfrac{1}{\frac{1}{3} + \frac{1}{4} + \frac{1}{5}}$ to 2 d.p.

8) $\dfrac{3.76 + 42.2}{1.60 + 0.86}$

9) $\dfrac{4.82 + 7.93}{-0.730 \times 6.92}$

10) $9.38(4.86 + 7.60 \times 1.89^3)$

11) $4.93^2 - 6.86^2$

12) $(4.93 + 6.86)(4.93 - 6.86)$

13) $\dfrac{1}{6.3^2 + 9.6^2}$

14) $\dfrac{3.864^2 + 9.62}{3.74 - 8.62^2}$

15) $\dfrac{9.5}{(6.4 \times 3.2) - (6.7 \times 0.9)}$

16) $1 - \dfrac{5.0}{3.6 + 7.49}$

17) $\frac{1}{6} - \frac{1}{5}(4.6)^2$

18) $\dfrac{6.4}{20.2}\left(3.94^2 - \dfrac{5.7 + 4.9}{6.7 - 3.2}\right)$

19) $\dfrac{3.64^3 + 5.6^2 - (1/0.085)}{9.76 + 3.4 - 2.9}$

20) $\dfrac{6.54(7.69 \times 10^{-5})}{0.643^2 - 79.3(3.21 \times 10^{-4})}$

Square root and 'pi' keys

$\boxed{\sqrt{}}$ gives the square root of any number in the display.

$\boxed{\pi}$ gives the numerical value of π to whatever accuracy the machine is designed.

EXAMPLE 1.19

The period, T seconds (the time for a complete swing), of a simple pendulum is given by the formula $T = 2\pi\sqrt{\dfrac{l}{g}}$ where l m is its length and g m/s^2 is the acceleration due to gravity.

Find the value of T if $l = 1.37$ m and $g = 9.81$ m/s^2.

Substituting the given values into the formula we have

$$T = 2\pi\sqrt{\dfrac{1.37}{9.81}}$$

The rough check gives: $T = 2 \times 3\sqrt{\frac{1}{9}} = 2 \times 3 \times \frac{1}{3} = 2$ s

The sequence of operations would be:

$$\boxed{AC}\boxed{1}\boxed{\cdot}\boxed{9}\boxed{7}\boxed{\div}\boxed{9}\boxed{\cdot}\boxed{8}\boxed{1}\boxed{=}\boxed{\sqrt{}}\boxed{\times}\boxed{2}\boxed{\times}\boxed{\pi}\boxed{=}$$

The display gives $\boxed{2.348\ 040\ 8}$

Thus the value of T is 2.35 seconds, correct to three significant figures.

The power key

$\boxed{x^y}$ gives the value of x to the index y.

EXAMPLE 1.20

The relationship between the luminosity, I, of a metal filament lamp and the voltage, V, is given by the equation $I = aV^4$ where a is a constant. Find the value of I if $a = 9 \times 10^{-7}$ and $V = 60$

Substituting the given values into the equation we have

$$I = (9 \times 10^{-7})60^4$$

The rough check gives

$$I = (10 \times 10^{-7})(6 \times 10)^4 = 10^{-6} \times 6^4 \times 10^4 = 10^{-2} \times 36 \times 36$$

and if we approximate by putting 30×40 instead of 36×36

$$\text{then} \quad I = 10^{-2} \times 30 \times 40 = 10^{-2} \times 1200 = 12$$

The sequence of operations would be:

$$\boxed{AC}\,\boxed{6}\,\boxed{0}\,\boxed{x^y}\,\boxed{4}\,\boxed{=}\,\boxed{\times}\,\boxed{9}\,\boxed{EXP}\,\boxed{7}\,\boxed{+/-}\,\boxed{=}$$

The display gives $\boxed{11.664}$

Thus the value of I is 11.7 correct to three significant figures.

EXAMPLE 1.21

The law of expansion of a gas is given by the expression $pV^{1.2} = k$ where p is the pressure, V is the volume, and k is a constant. Find the value of k if $p = 0.8 \times 10^6$ and $V = 0.2$

Substituting the given values into the formula we have

$$k = (0.8 \times 10^6) \times 0.2^{1.2}$$

The rough check gives

$$k = 1 \times 10^6 \times \left(\frac{2}{10}\right)^{1.2} = 10^6 \times \frac{2^{1.2}}{10^{1.2}} = 10^6 \times \frac{3}{30} = 0.1 \times 10^6$$

Since it is difficult to assess the approximate value of a decimal number to an index, it becomes simpler to express the decimal number as a fraction using whole numbers. In this case it is convenient to express 0.2 as $\frac{2}{10}$. We can guess the rough value of $2^{1.2}$, since we know that $2^1 = 2$ and $2^2 = 4$. Similarly we judge the value of $10^{1.2}$ as being between $10^1 = 10$ and $10^2 = 100$. The more practice you have in doing calculations of this type, the more accurate your guess will be.

The sequence of operations would be:

$$\boxed{AC}\,\boxed{0}\,\boxed{\cdot}\,\boxed{2}\,\boxed{x^y}\,\boxed{1}\,\boxed{\cdot}\,\boxed{2}\,\boxed{=}\,\boxed{\times}\,\boxed{0}\,\boxed{\cdot}\,\boxed{8}\,\boxed{EXP}\,\boxed{6}\,\boxed{=}$$

The display gives $\boxed{115\,964.74}$

Thus the required value of k is 116 000 or 0.116×10^6 correct to three significant figures.

Nesting (or stacking) a polynomial

A polynomial in x is an expression containing a sum of terms, each term being a power of x.

A typical polynomial is

$$ax^4 + bx^3 + cx^2 + dx + e$$

where a, b, c, d and e are constants.

The polynomial may be factorised successively as follows:

$$
\begin{aligned}
&= \quad ax^4 + bx^3 + cx^2 + dx + e \\
&= (ax + b)x^3 + cx^2 + dx + e \\
&= \{(ax + b)x + c\}x^2 + dx + e \\
&= [\{(ax + b)x + c\}x + d]x + e
\end{aligned}
$$

The opening bracket symbols [{(are usually omitted and all the remaining closure brackets are written in the same form. Thus the polynomial looks like:

$$
ax + b)x + c)x + d)x + e
$$

This is called the nested (or stacked) form of the polynomial. When evaluating its value we must always work from *left* to *right*.

EXAMPLE 1.22

Find the value of $y = 5x^3 - 7x^2 + 8x - 5$ when $x = 3.32$

The nested form gives $y = 5x - 7)x + 8)x - 5$

When $x = 3.32$ then $y = 5 \times 3.32 - 7)3.32 + 8)3.32 - 5$

Working from *left* to *right* the sequence of operation is:

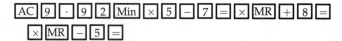

The display gives $\boxed{127.375\ 04}$

Thus $y = 127$ correct to three significant figures.

Note that if the polynomial is not nested it would be necessary to evaluate

$$
5(3.32)^3 - 7(3.32)^2 + 8(3.32) - 5
$$

Try this for yourself and you will appreciate the advantage of nesting.

Trigonometrical functions

Use of [sin], [cos], [tan] and [inv] keys.

EXAMPLE 1.23

$$
\text{Find angle } A \text{ if } \quad \sin A = \frac{3.68 \sin 42^\circ}{5.26}
$$

Rough check: It is always difficult to find an approximate answer for a calculation involving trigonometrical functions. However, we may use a 'backwards substitution' method which is carried out after an answer has been obtained.

On most caculators there is a mode selection which may be set at either RAD (radians), DEG (degrees) or GRAD (grades—a grade being one-hundreth of a right angle, used more on the continent). In this example the angles are in degrees and so we set the mode to the DEG position.

A sequence of operations commencing with

$$\boxed{\text{AC}}\,\boxed{9}\,\boxed{.}\,\boxed{6}\,\boxed{8}\,\boxed{\times}\,\boxed{4}\,\boxed{2}\,\boxed{\text{sin}}\,\boxed{=}$$

may not function correctly as it merely gives the value of sin 42°. Try it on your own machine. We must, therefore, alter the order in which the operations are carried out. Thus the sequence used should be:

$$\boxed{\text{AC}}\,\boxed{4}\,\boxed{2}\,\boxed{\text{sin}}\,\boxed{\times}\,\boxed{9}\,\boxed{.}\,\boxed{6}\,\boxed{8}\,\boxed{\div}\,\boxed{5}\,\boxed{.}\,\boxed{2}\,\boxed{6}\,\boxed{=}\,\boxed{\text{inv}}\,\boxed{\text{sin}}$$

The display gives $\boxed{27.913\ 433}$

Thus angle $A = 27.9°$ correct to three significant figures.

Answer check: As the result of our calculations we have

$$\sin 27.9° = \frac{3.68 \sin 42°}{5.26}$$

Now we may rearrange this expression to give

$$\sin 42° = \frac{5.26 \sin 27.9°}{3.68}$$

Thus if we find the value of the right-hand side of this new expression, it should give the value of sin 42° if the value of angle $A = 27.9°$ is correct.

The sequence of operations is similar to that given for the original calculation. Try it for yourself and thus check the result.

Exercise 1.7

Evaluate the exressions in Questions 1–8 giving the answers correct to three significant figures.

1) 2.32^4 2) 1.52^6 3) 0.523^5 4) 7.9^{-2}

5) 4.59^{-3} 6) 0.321^{-4} 7) $12.1^{1.5}$ 8) $6.83^{2.32}$

Evaluate the expressions in Questions 9–16 giving the answers correct to two decimal places.

9) $0.879^{3.1}$ 10) $5.56^{0.62}$ 11) $14.7^{0.347}$ 12) $3.9^{-0.5}$

13) $6.64^{3/2}$ 14) $13.6^{2/5}$ 15) $1.23^{7/3}$ 16) $0.334^{3/5}$

In Questions 17–30 the accuracy to which you give the answer will depend on the given figures in each set of data.

17) Evaluate $4\pi r^2$ when $r = 6.1$

18) Evaluate $5\pi(R^2 - r^2)$ when $R = 1.32$ and $r = 1.24$

19) In a beam the stress, σ, due to bending is given by the expression

$\sigma = \dfrac{My}{I}$. Find σ if $M = 12 \times 10^6$, $y = 60$ and $I = 11.5 \times 10^6$

20) The polar second moment of area, J, of a hollow shaft is given by the equation $J = \dfrac{\pi}{32}(D^4 - d^4)$. Find J if $D = 220$ and $d = 140$

21) The velocity, v, of a body performing simple harmonic motion is given by the expression $v = \omega\sqrt{A^2 - x^2}$. Find v if $\omega = 20.9$, $A = 0.060$, and $x = 0.020$

22) The natural frequency of oscillation, f, of a mass, m, supported by a spring of stiffness, λ, is given by the formula $f = \dfrac{1}{2\pi}\sqrt{\dfrac{\lambda}{m}}$. Find f if $\lambda = 5000$ and $m = 1.5$

23) Find the value of a if $a = \dfrac{80.6 \sin 55°}{\sin 70°}$

24) If $\cos C = \dfrac{a^2 + b^2 - c^2}{2ab}$, find the value of the angle C when $a = 19.37\,\text{mm}$, $b = 26.42\,\text{mm}$ and $c = 22.31\,\text{mm}$

25) Use the method of nesting to evaluate the following:
   a) $5x^2 + 4x - 15$ when $x = 3.8$ correct to 1 d.p.
   b) $7x^3 + 3x^2 - 7x + 5$ when $x = 0.35$ correct to 2 d.p.
   c) $x^4 + 3x^3 - 8x^2 + x - 3$ when $x = 3.75$ correct to 3 s.f.
   d) $x^3 + 5x^2 - 6x + 9$ when $x = 1.2$ correct to 1 d.p.

Number systems

So far, in mathematics we have only used the ten digit, or denary, system. You have become familiar with this by counting using the ten digits 0 to 9. However, this is not the only system used in engineering. Occasionally the octal system is used, which is based on 8, whilst hexadecimal (based on 16) and binary (based on 2) systems are invariably used in computers and associated technology.

The denary system

Let us first consider the ordinary decimal (or denary) system which uses ten digits 0 to 9, and a number base of ten. For example:

$$2975 = 2000 + 900 + 70 + 5$$
$$= 2 \times 1000 + 9 \times 100 + 7 \times 10 + 5 \times 1$$
$$= 2 \times 10^3 + 9 \times 10^2 + 7 \times 10^1 + 5 \times 10^0$$

Another way of showing this is a series of divisions:

$$
\begin{aligned}
2975 \div 10 &= 297 \text{ remainder } 5 \text{ — units} \\
297 \div 10 &= 29 \text{ remainder } 7 \text{ — tens} \\
29 \div 10 &= 2 \text{ remainder } 9 \text{ — hundreds} \\
2 \div 10 &= 0 \text{ remainder } 2 \text{ — thousands}
\end{aligned}
$$

$$2 \quad 9 \quad 7 \quad 5$$

Why did we choose a ten based system? Probably because we had a total of ten fingers and thumbs and these were our original calculating machine!

The octal system

Suppose now that the inventor of a number system had lost both his thumbs in a hunting accident—so his counting would have to be done on eight fingers. The octal system has a number base of eight, and uses eight digits 0 to 7.

So let us use the series of divisions method to convert 2975 denary to octal.

$$2975 \div 8 = 371 \text{ remainder } 7 \text{ — units}$$
$$371 \div 8 = 46 \text{ remainder } 3 \text{ — eights}$$
$$46 \div 8 = 5 \text{ remainder } 6 \text{ — } (8 \times 8)\text{s}$$
$$5 \div 8 = 0 \text{ remainder } 5 \text{ — } (8 \times 8 \times 8)\text{s}$$

$$5 \quad 6 \quad 3 \quad 7$$

So the octal number 5637 represents $\Big\} 5 \times 8^3 + 6 \times 8^2 + 3 \times 8^1 + 7 \times 8^0$

Thus $\qquad\qquad\qquad 2975 \text{ (denary)} = 5637 \text{ (octal)}$

Or using notation $\qquad\qquad 2975_{10} = 5637_8$

EXAMPLE 1.24
Convert 71 293 to octal.

$$71\,293 \div 8 = 8911 \text{ remainder } 5 \quad \text{units}$$
$$8911 \div 8 = 1113 \text{ remainder } 7 \quad \text{eights}$$
$$1113 \div 8 = 139 \text{ remainder } 1 \quad (8^2)\text{s}$$
$$139 \div 8 = 17 \text{ remainder } 3 \quad (8^3)\text{s}$$
$$17 \div 8 = 2 \text{ remainder } 1 \quad (8^4)\text{s}$$
$$2 \div 8 = 0 \text{ remainder } 2 \quad (8^5)\text{s}$$

Thus $\quad 71\,293_{10} = 2 \times 8^5 + 1 \times 8^4 + 3 \times 8^3 + 1 \times 8^2 + 7 \times 8^1 + 5$

or $\quad 71\,293_{10} = 213\,175_8$

EXAMPLE 1.25
Convert $174\,023_8$ to denary.

$$174\,023_8 = 1 \times 8^5 + 7 \times 8^4 + 4 \times 8^3 + 0 \times 8^2 + 2 \times 8^1 + 3$$
$$= 32\,768 + 28\,672 + 2048 + 0 + 16 + 3$$
$$= 63\,507$$

so $\quad 174\,023_8 = 63\,507_{10}$

Octal arithmetic

Counting in octal: 0, 1, 2, 3, 4, 5, 6, 7, ?

There is no 8 so we continue with 10 meaning 1 eight and no units:

$$0, 1, 2, 3, 4, 5, 6, 7, 10, 11, 12, 13, 14, 15, 16, 17,$$
$$20, 21, 22, 23, \ldots 75, 76, 77, \quad ?$$

There is no 78 or 80 and we move to 100 which means one $8^2 = 64$ and no eights and no units, giving 76, 77, 100, 101, 102, ...

Adding : 1567_8

$+\ \ 154_8$

$11 \leftarrow$ carries

$\overline{1743_8}$

Working from the R.H. column to the left, then

$7 + 4 = 13_8$ giving 3 carry 1

$6 + 5 + 1 = 14_8$ giving 4 carry 1

$5 + 1 + 1 = 7_8$ giving 7 etc.

The binary system

This is a number system using a base of two—this means that only two digits are needed, namely 0 and 1. You will immediately note, no doubt, why binary numbers are favoured in computer logic since this uses only two constants, namely 0 and 1.

Let us use the series of divisions method to convert 45 to binary:

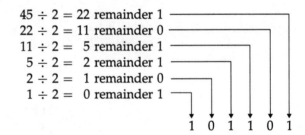

$$45 \div 2 = 22 \text{ remainder } 1$$
$$22 \div 2 = 11 \text{ remainder } 0$$
$$11 \div 2 = \ \ 5 \text{ remainder } 1$$
$$5 \div 2 = \ \ 2 \text{ remainder } 1$$
$$2 \div 2 = \ \ 1 \text{ remainder } 0$$
$$1 \div 2 = \ \ 0 \text{ remainder } 1$$

$$1 \quad 0 \quad 1 \quad 1 \quad 0 \quad 1$$

So the binary number 101101 represents:

$$1 \times 2^5 + 0 \times 2^4 + 1 \times 2^3 + 1 \times 2^2 + 0 \times 2^1 + 1 \times 2^0$$

or $\ 1 \times 32 + 0 \times 16 + 1 \times 8 + 1 \times 4 + 0 \times 2 + 1 \times 1$

or $\ 32 + 8 + 4 + 1 = 45$

Thus $\qquad\qquad\qquad$ 45 denary $= 101101$ binary

or $\qquad\qquad\qquad\qquad 45_{10} = 101101_2$

Bicimals

In the denary system figures to the right of the decimal point are called decimals. In the binary system figures to the right of the bicimal point are called bicimals.

So the binary number 101.01101 represents:

$$1 \times 2^2 + 0 \times 2^1 + 1 \times 2^0 + 0 \times 2^{-1} + 1 \times 2^{-2} + 1 \times 2^{-3} + 0 \times 2^{-4}$$
$$+ 1 \times 2^{-5}$$

or $\quad 1 \times 4 + 0 \times 2 + 1 \times 1 + 0 \times \frac{1}{2} + 1 \times \frac{1}{4} + 1 \times \frac{1}{8} + 0 \times \frac{1}{16} + 1 \times \frac{1}{32}$

or $\quad 4 + 1 + \frac{1}{4} + \frac{1}{8} + \frac{1}{32}$

or $\quad 4 + 1 + 0.25 + 0.125 + 0.03125$

or $\quad 5.40625$

Thus $\qquad 101.01101$ binary $= 5.406\,25$ denary

or $\qquad 101.01101_2 = 5.406\,25_{10}$

To convert decimal $0.406\,25$ to bicimal we use a series of multiplications:

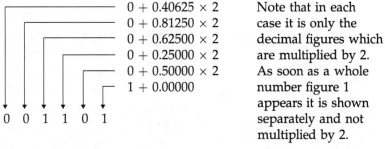

$0 + 0.40625 \times 2$	Note that in each
$0 + 0.81250 \times 2$	case it is only the
$0 + 0.62500 \times 2$	decimal figures which
$0 + 0.25000 \times 2$	are multiplied by 2.
$0 + 0.50000 \times 2$	As soon as a whole
$1 + 0.00000$	number figure 1

appears it is shown separately and not multiplied by 2.

Thus $\qquad 0.406\,25_{10} = 0.01101_2$

Binary arithmetic

An understanding of binary arithmetic is useful since most computer circuits perform calculations using binary although they may not use the same methods as shown here. We shall show how to add and multiply in binary.

EXAMPLE 1.26
Add 10101_2 and 1111_2.

Working from the RH column to the left then:

$$1 + 1 = 10 \text{ giving } 0 \text{ carry } 1$$
$$0 + 1 + 1 = 10 \text{ giving } 0 \text{ carry } 1$$
$$1 + 1 + 1 = 11 \text{ giving } 1 \text{ carry } 1$$
and so on.

```
  1 0 1 0 1
+   1 1 1 1
  1 1 1 1 ← carries
  ─────────
1 0 0 1 0 0₂
```

EXAMPLE 1.27
Multiply 10011_2 by 1101_2.

```
        1 0 0 1 1
  ×       1 1 0 1
  ───────────────
        1 0 0 1 1   (a)
        0 0 0 0 0   (b)
      1 0 0 1 1     (c)
    1 0 0 1 1       (d)
  ───────────────
```

Sum of lines a, b, c and d.

The computer would not add the four lines a, b, c and d in one go, but would probably proceed in stages:

$$
\begin{array}{ll}
1\,0\,0\,1\,1 & \text{(a)} \\
+\,0\,0\,0\,0\,0 & \text{(b)} \\
\hline
1\,0\,0\,1\,1 & \text{(a)} + \text{(b)} \\
+\,1\,0\,0\,1\,1 & \text{(c)} \\
\hline
1\,0\,1\,1\,1\,1\,1 & \text{(a)} + \text{(b)} + \text{(c)} \\
+\,1\,0\,0\,1\,1 & \text{(d)} \\
1\,1 & \leftarrow \text{carries} \\
\hline
\end{array}
$$

$$1\,1\,1\,1\,0\,1\,1\,1\,1_2 \qquad \text{Answer: sum of a, b, c and d.}$$

Exercise 1.8

1) Convert to denary
   a) 63_8 b) 217_8 c) 463_8

2) Convert to octal
   a) 29_{10} b) 243_{10} c) 1796_{10}

3) Find the sum of
   a) 637_8 and 56_8 b) 4360_8 and 1234_8

4) Check the results of Question 3 by converting each number to denary, adding, and then converting back to octal.

5) Convert to binary
   a) 23_{10} b) 125_{10} c) 97_{10}

6) Convert to denary
   a) 10110   b) 111001   c) 1011010

7) Convert to denary
   a) 0.1101   b) 0.0111   c) 0.0011

8) Convert to binary
   a) $\frac{3}{8}$ b) $\frac{5}{16}$ c) $\frac{7}{8}$

9) Convert to binary and correct to seven bicimal places:
   a) 0.169   b) 18.467   c) 108.710

   Hint: In (b) and (c) convert the whole number parts separately from the decimal parts and then combine the results for the final answer.

10) Add the following binary numbers:
    a) $1011 + 11$ b) $11011 + 1011$
    c) $10111 + 11010 + 111$

11) Multiply the following binary numbers:
    a) 101×111 b) 1011×1010 c) 11011×1101

Perimeters, areas and volumes

Areas and perimeters

Most complicated forms may be divided into simple shapes such as rectangles, triangles and circles. This means, of course, that we must be familiar with the areas and perimeters of these basic figures. Engineers constantly use areas: for instance, stress is calculated from force and area, as is pressure. Another example is electrical conductivity, which depends on the cross-sectional area of the conductor through which the current flows.

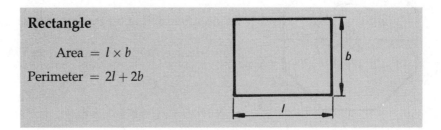

Rectangle

$$\text{Area} = l \times b$$

$$\text{Perimeter} = 2l + 2b$$

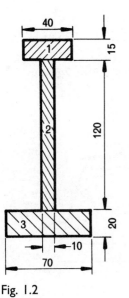

Fig. 1.2

EXAMPLE 1.28

Find the area of the section shown in Fig. 1.2.

The section can be split up into three rectangles as shown. The total area can be found by calculating the areas of the three rectangles separately and then adding these together. Thus,

$$\text{Area of rectangle } 1 = 15 \times 40 = 600 \text{ mm}^2$$

$$\text{Area of rectangle } 2 = 10 \times 120 = 1200 \text{ mm}^2$$

$$\text{Area of rectangle } 3 = 20 \times 70 = 1400 \text{ mm}^2$$

$$\text{Total area of section} = 600 + 1200 + 1400 = 3200 \text{ mm}^2$$

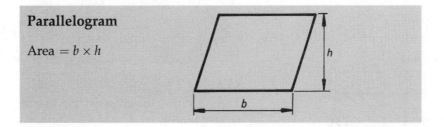

Parallelogram

$$\text{Area} = b \times h$$

EXAMPLE 1.29

Find the area of the parallelogram shown in Fig. 1.3.

The first step is to find the vertical height h.

In $\triangle BCE$, $\quad h = BC \times \sin 60° = 3 \times 0.866 = 2.598$

$$\text{Area of parallelogram} = \text{Base} \times \text{Vertical height}$$

$$= 5 \times 2.598$$

$$= 13.0 \text{ m}^2$$

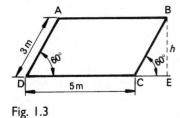

Fig. 1.3

Triangle

Area $= \frac{1}{2} \times b \times h$

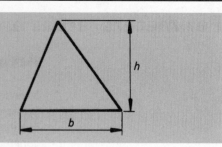

EXAMPLE 1.30

The inner shape of a pattern in a large hall is a regular octagon (8-sided polygon) which is 5 m across flats (Fig. 1.4). Find its area.

Angle subtended at centre by a side of the octagon $= \dfrac{360°}{8} = 45°$.

Now triangle AOB is isosceles, since OA = OB.

$$\therefore \qquad \angle AOC = \frac{45°}{2} = 22°30'$$

But $\qquad OC = \dfrac{5}{2} = 2.5 \text{ m}$

Also $\qquad \dfrac{AC}{OC} = \tan 22°30'$

$\therefore \qquad AC = OC \times \tan 22°30' = 2.5 \times 0.4142 = 1.035 \text{ m}$

Thus $\qquad$ Area of $\triangle AOB = AC \times OC = 1.035 \times 2.5 = 2.588 \text{ m}^2$

$\therefore \qquad$ Area of octagon $= 2.588 \times 8 = 20.7 \text{ m}^2$

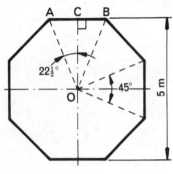

Fig. 1.4

Trapezium (or trapezoid)

Area $= \frac{1}{2} \times h \times (a + b)$

EXAMPLE 1.31

Fig. 1.5 shows the cross-section of a retaining wall. Calculate its cross-sectional area.

Since the section is a trapezium:

$$
\begin{aligned}
\text{Area} \quad &= \tfrac{1}{2} \times h \times (a + b) \\
&= \tfrac{1}{2} \times 6 \times (2 + 3) \\
&= \tfrac{1}{2} \times 6 \times 5 \\
&= 15 \text{ m}^3
\end{aligned}
$$

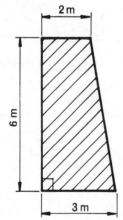

Fig. 1.5

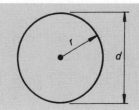

Circle

$$\text{Area} = \pi r^2 = \frac{\pi d^2}{4}$$

$$\text{Circumference} = 2\pi r = \pi d$$

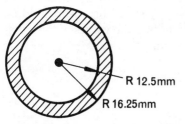

R 12.5mm

R 16.25mm

Fig. 1.6

EXAMPLE 1.32
A pipe has an outside diameter of 32.5 mm and an inside diameter of 25 mm. Calculate the cross-sectional area of the shaft (Fig. 1.6).

Area of cross-section = Area of outside circle − Area of inside circle

$$= \pi \times 16.25^2 - \pi \times 12.5^2 = 338 \text{ mm}^2$$

Exercise 1.9

1) The area of a metal plate is 220 mm^2. If its width is 25 mm, find its length.

2) A sheet metal plate has a length of 147.5 mm and a width of 86.5 mm. Find its area in m^2.

3) Find the areas of the cross-sections of the aluminium extrusions shown in Fig. 1.7.

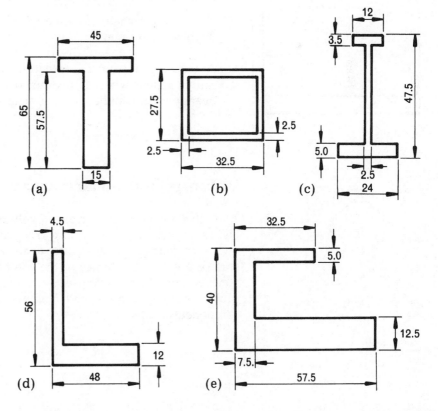

Fig. 1.7

4) Whilst working on the specification for a new kitchen appliance, a draughtsman has the problem of finding the values of two areas:

a) The area of a parallelogram if two adjacent sides measure 112.5 mm and 10.5 mm, and the angle between them is 49°.

b) The area of a trapezium whose parallel sides are 75 mm and 82 mm long, respectively, and whose vertical height is 39 mm.

5) In the course of working out the cost of some bright drawn steel bars, a cost engineer has to find their masses. This means finding their volumes which, in turn, means finding the values of the cross-sectional areas which have the following shapes:

a regular hexagon

a) which is 40 mm wide across flats,

b) which has sides 50 mm long

a regular octagon

c) which is 20 mm wide across flats

d) which has sides 20 mm long.

6) If the area of cross-section of a circular shaft is 700 mm², find its diameter.

7) A company which manufactures reciprocating engines requires the value of the shaded area in Fig. 1.8 as this is the cross-sectional area of flow of a liquid coolant channel.

8) A hollow shaft has a cross-sectional area of 868 mm². If its inside diameter is 7.5 mm, calculate its outside diameter.

9) Find the area of the blank shown in Fig. 1.9.

10) How many revolutions will a wheel make in travelling 2 km if its diameter is 700 mm?

11) A rectangular piece of insulating material is required to wrap round a pipe which is 560 mm diameter. Allowing 150 mm for overlap, calculate the width of material required.

Fig. 1.8

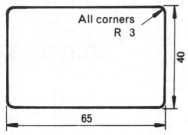

All corners R 3

40

65

Fig. 1.9

Volumes and surface areas

As with perimeters and areas, so volumes and surface areas can be just as important in engineering. The same technique of subdividing awkward solids is used, so once again we must be familiar with the surface areas and volumes of the component shapes, e.g. cylinders, spheres etc. Areas are used in heat radiation and conduction problems for example, and volumes in finding the masses of components.

> ## Prism
>
> A prism is the name given to a solid with a uniform cross-section and parallel end faces.
>
> Volume = Cross-sectional area × Length of solid
>
> Surface area = Longitudinal surface + Ends
>
> i.e. (Perimeter of cross-section × Length of solid)
>
> + (Total area of ends)

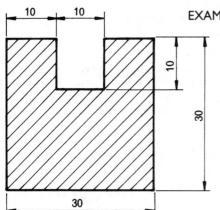

Fig. 1.10

EXAMPLE 1.33

A piece of timber has the cross-section shown in Fig. 1.10. If its length is 300 mm, find its volume and total surface area.

$$\text{Area of cross-section} = (30 \times 30) - (10 \times 10) = 800 \text{ mm}^2$$
$$\text{Volume} = (\text{Area of cross-section}) \times (\text{Length})$$
$$= 800 \times 300 = 240\,000 \text{ mm}^3$$
$$\text{Perimeter of cross-section} = (3 \times 30) + (5 \times 10) = 140 \text{ mm}$$
$$\text{Total surface area} = (\text{Perimeter of cross-section} \times \text{Length})$$
$$+ (\text{Area of ends})$$
$$= (140 \times 300) + (2 \times 800) = 43\,600 \text{ mm}^2$$

EXAMPLE 1.34

A steel section has the cross-section shown in Fig. 1.11. If it is 9 m long, calculate its volume, total surface area and mass.

To find the volume

$$\text{Area of cross-section} = \tfrac{1}{2} \times \pi \times 75^2 + 100 \times 150 = 23\,840 \text{ mm}^2$$
$$= \frac{23\,840}{(1000)^2} = 0.023\,84 \text{ m}^2$$
$$\therefore \qquad \text{Volume of solid} = 0.023\,84 \times 9 = 0.215 \text{ m}^3$$

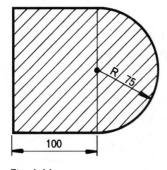

Fig. 1.11

To find the surface area

$$\text{Perimeter of cross-section} = \pi \times 75 + 2 \times 100 + 150$$
$$= 585.5 \text{ mm} = \frac{585.5}{1000} = 0.5855 \text{ m}$$
$$\text{Lateral surface area} = 0.5855 \times 9 = 5.270 \text{ m}^2$$
$$\text{Surface area of ends} = 2 \times 0.024 = 0.048 \text{ m}^2$$
$$\therefore \qquad \text{Total surface area} = 5.27 + 0.05 = 5.32 \text{ m}^2$$

To find the mass

If we use the density of steel of 7900 kg/m^3 and

we know that $\qquad\qquad$ mass = volume × density

then $\qquad\qquad\qquad$ mass = $0.215 \times 7900 \ \cancel{\text{m}^3} \times \dfrac{\text{kg}}{\cancel{\text{m}^3}}$

$$= 1700 \text{ kg correct to 3 s.f.}$$

Cylinder

Volume $= \pi r^2 h$

Surface area $= 2\pi rh + 2\pi r^2 = 2\pi r(h + r)$

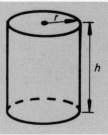

EXAMPLE 1.35

A cylindrical can holds 18 litres of petrol. Find the depth of the petrol if the can has a diameter of 600 mm.

Now   18 litres $= 18 \times 10^6$ mm^3

and if the depth of the petrol is h mm

then $\qquad$ Volume of petrol $= \pi \, (\text{Radius})^2 \times h$

$\therefore \qquad\qquad\qquad 18 \times 10^6 = \pi \times 300^2 \times h$

$\therefore \qquad\qquad\qquad\qquad h = \dfrac{18\,000\,000}{\pi \times 90\,000} = 63.7$ mm

EXAMPLE 1.36

A metal bar of length 200 mm and diameter 75 mm is melted down and cast into washers 2.5 mm thick with an internal diameter of 12.5 mm and external diameter 25 mm. Calculate the number of washers obtained assuming no loss of metal.

$$\begin{aligned}
\text{Volume of original bar of metal} &= \pi \times 37.5^2 \times 200 \\
&= 883\,500 \text{ mm}^3 \\
\text{Volume of one washer} &= \pi \times (12.5^2 - 6.25^2) \times 2.5 \\
&= \pi \times 117.2 \times 2.5 \\
&= 920.4 \text{ mm}^3 \\
\text{Number of washers obtained} &= \frac{883\,500}{920.4} = 960
\end{aligned}$$

Cone

Volume $= \frac{1}{3}\pi r^2 h$

(h is the vertical height)

Curved surface area $= \pi r l$

(l is the slant length)

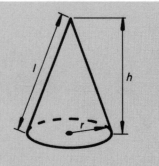

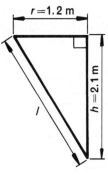

Fig. 1.12

EXAMPLE 1.37

A hopper is in the form of an inverted cone. It has a maximum internal diameter of 2.4 m and an internal height of 2.1 m.

a) If the hopper is to be lined with lead, calculate the area of lead required.

b) Determine the capacity of the hopper before lining.

a) The slant height may be found by using Pythagoras' theorem on the triangle shown in Fig. 1.12.

$$l^2 = r^2 + h^2 = 1.2^2 + 2.1^2 = 5.85$$
$$l = \sqrt{5.85} = 2.42 \text{ m}$$
$$\text{Surface area} = \pi r l = \pi \times 1.2 \times 2.42 = 9.12 \text{ m}^2$$

Hence the area of lead required is $9.12\,\text{m}^2$

b) Volume $= \frac{1}{3}\pi r^2 h = \frac{1}{3} \times \pi \times 1.2^2 \times 2.1 = 3.17 \text{ m}^3$

Hence the capacity of the hopper is $3.17\,\text{m}^3$

Frustum of a cone

A *frustum* is the portion of a cone or pyramid between the base and a horizontal slice which removes the pointed portion.

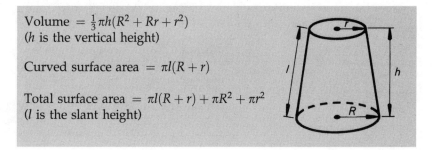

Volume $= \frac{1}{3}\pi h(R^2 + Rr + r^2)$
(h is the vertical height)

Curved surface area $= \pi l(R + r)$

Total surface area $= \pi l(R + r) + \pi R^2 + \pi r^2$
(l is the slant height)

EXAMPLE 1.38

A concrete column is shaped like a frustum of a cone. It is 2.8 m high. The radius at the top is 0.6 m and the base radius is 0.9 m. Calculate the volume of concrete in the column.

$$\begin{aligned}
\text{Volume} &= \tfrac{1}{3}\pi h(R^2 + Rr + r^2) \\
&= \tfrac{1}{3} \times \pi \times 2.8 \times (0.9^2 + 0.9 \times 0.6 + 0.6^2) \\
&= \tfrac{1}{3} \times \pi \times 2.8 \times 1.71 = 5.01 \text{ m}^3
\end{aligned}$$

EXAMPLE 1.39

The bowl shown in Fig. 1.13 is made from sheet steel and has an open top. Calculate the total cost of painting the vessel (inside and outside) at a cost of 1 p per 10 000 mm².

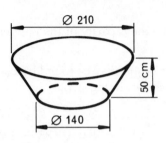

Fig. 1.13

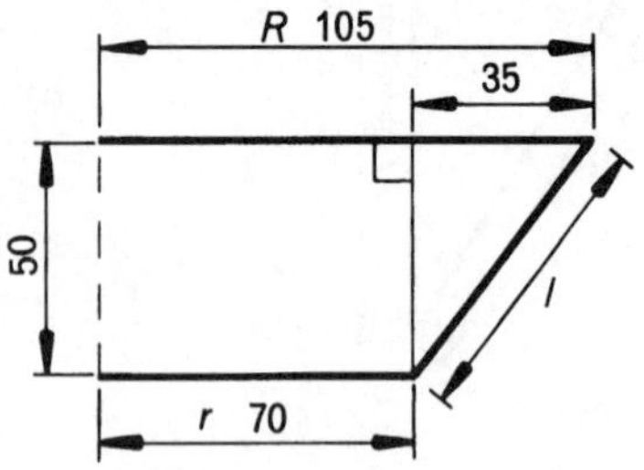

Fig. 1.14

Fig. 1.14 shows a half section of the bowl. Using Pythagoras' theorem on the right-angled triangle

$$l^2 = 50^2 + 35^2$$
$$\therefore \quad l = 61.0 \text{ mm}$$

Now the required total surface area, i.e. inside and outside

$$= 2\{\text{Curved surface area}\} + 2(\text{Base area})$$
$$= 2\{\pi l(R + r)\} + 2(\pi r^2)$$
$$= 2\{\pi(61)(105 + 70)\} + 2(\pi 70^2) = 97\ 800 \text{ mm}^2$$

At 1 p per 10 000 mm² total cost $= \dfrac{97\ 800}{10\ 000} = 9.78$ p

---

Sphere

Volume $= \frac{4}{3}\pi r^3$

Surface area $= 4\pi r^2$

EXAMPLE 1.40

A metal end cap is in the shape of a hemisphere. Its internal and external diameters are 48 mm and 50 mm, respectively.

a) Calculate the volume of metal used in its construction.

b) If the inside is to be painted, calculate the area to be covered.

a) If R is the outside radius, r the inside radius and V the volume, then

$$V = \frac{1}{2}(\frac{4}{3}\pi R^3 - \frac{4}{3}\pi r^3)$$
$$= \frac{1}{2} \times \frac{4}{3} \times \pi \times (R^3 - r^3)$$
$$= \frac{2}{3} \times \pi \times (25^3 - 24^3) = 3770 \text{ mm}^3$$

b) $\qquad$ Area of inside of dome $= \frac{1}{2} \times 4\pi r^2$
$$= \frac{1}{2} \times 4 \times \pi \times 24^2 = 3620 \text{ mm}^2$$

Pyramid

Volume $= \frac{1}{3}Ah$

Surface area $=$ Sum of the areas of
the triangles forming
the sides plus the
area of the base

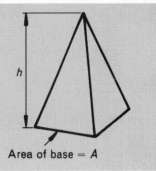

Area of base $= A$

EXAMPLE 1.41

Find the volume and total surface area of a symmetrical pyramid whose base is a rectangle $7\,\text{m} \times 4\,\text{m}$ and whose height is $10\,\text{m}$.

$$\text{Base area} = 7 \times 4 = 28 \text{ m}^2$$
$$\text{Height} = 10 \text{ m}$$
$$\text{Volume} = \frac{1}{3}Ah$$
$$= \frac{1}{3} \times 28 \times 10$$
$$= 93.3 \text{ m}^3$$

To find surface area

From Fig. 1.15 the surface area consists of two sets of equal triangles (that is $\triangle ABC$ and $\triangle ADE$, and also $\triangle ABE$ and $\triangle ACD$) together with the base BCDE. To find the area of $\triangle ABC$ we must find the slant height AH. From the apex, A, drop a perpendicular AG on to the base and draw GH perpendicular to BC. H is then the mid-point of BC.

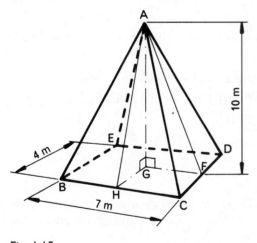

Fig. 1.15

In $\triangle AHG$, $\angle AGH = 90°$ and, by Pythagoras' theorem

$$AH^2 = AG^2 + HG^2 = 10^2 + 2^2 = 104$$
$$\therefore \qquad AH = \sqrt{104} = 10.2 \text{ m}$$
$$\therefore \quad \text{Area of } \triangle ABC = \frac{1}{2} \times \text{Base} \times \text{Height}$$
$$= \frac{1}{2} \times 7 \times 10.2$$
$$= 35.7 \text{ m}^2$$

Similarly, to find the area of $\triangle ACD$ we must find the slant height AF. Draw GF, F being the mid-point of CD. Then in $\triangle AGF$, $\angle AGF = 90°$ and by Pythagoras' theorem,

$$AF^2 = AG^2 + GF^2 = 10^2 + 3.5^2 = 112.3$$
$$\therefore \qquad AF = \sqrt{112.3} = 10.6 \text{ m}$$
$$\therefore \quad \text{Area of } \triangle ACD = \frac{1}{2} \times \text{Base} \times \text{Height}$$
$$= \frac{1}{2} \times 4 \times 10.6 = 21.2 \text{ m}^2$$
$$\therefore \text{Total surface area} = (2 \times 35.7) + (2 \times 21.2) + (7 \times 4)$$
$$= 142 \text{ m}^2$$

Frustum of a pyramid

Volume $= \frac{1}{3}h(A + \sqrt{Aa} + a)$

Surface area = Sum of the areas of the trapeziums forming the sides plus the areas of the top and base

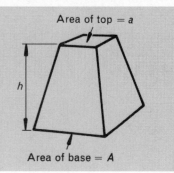

Area of top $= a$

h

Area of base $= A$

EXAMPLE 1.42

A casting has a length of 2 m and its cross-section is a regular hexagon. The casting tapers uniformly along its length, the hexagon having a side of 200 mm at one end and 100 mm at the other. Calculate the volume of the casting.

In Fig. 1.16 the area of the hexagon of 200 mm side $= 6 \times$ Area $\triangle ABO$

In $\triangle AOC$, $\angle AOC = 30°$, $AC = 100$ mm

and $\quad \tan\angle AOC = \dfrac{AC}{OC}$

giving $\quad OC = \dfrac{AC}{\tan\angle AOC°} = \dfrac{100}{\tan 30°} = 173$ mm

$\therefore$ Area of $\triangle ABO = \frac{1}{2} \times 200 \times 173$

$\therefore$ Area of hexagon $= 6 \times \frac{1}{2} \times 200 \times 173 = 103\,800$ mm^2

The area of the hexagon of 100 mm side can be found in the same way. It is 25 950 mm^2.

The casting is a frustum of a pyramid with $A = 103\,800$, $a = 25\,950$ and $h = 2000$

$\therefore \qquad$ Volume $= \frac{1}{3}h(A + \sqrt{Aa} + a)$

$= \frac{1}{3} \times 2000 \times (103\,800 + \sqrt{103\,800 \times 25\,950} + 25\,950)$

$= 1.21 \times 10^8$ mm^3

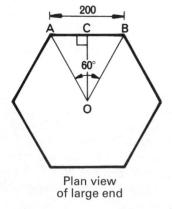

200

A   C   B

60°

O

Plan view
of large end

Fig. 1.16

Exercise 1.10

1) A block of lead 0.15 m by 0.1 m by 0.075 m is hammered out to make a square sheet 10 mm thick. What are the dimensions of the square?

2) Calculate the volume of metal in a pipe which has a bore of 50 mm, a thickness of 8 mm and a length of 6 m.

3) A hot-water cylinder whose length is 1.12 m is to hold 200 litres. Find the diameter of the cylinder in millimetres.

4) Calculate the heating surface (in square metres) of a steam pipe whose external diameter is 60 mm and whose length is 8 m.

5) A metal bucket is 400 mm deep. It is 300 mm diameter at the top and 200 mm diameter at the bottom. Calculate the number of litres of water that the bucket will hold assuming that it is a frustum of a cone.

6) It is required to replace two pipes with bores of 28 mm and 70 mm respectively with a single pipe which has the same area of flow. Find the bore of this single pipe.

7) A metal container shaped like a pyramid has a square base of side 2 m and a height of 4 m. Calculate its volume and its total surface area.

8) A column is a regular octagon (8-sided polygon) in cross-section. It is 460 mm across flats at the base and it tapers uniformly to 300 mm across flats at the top. If it is 3.6 m high, calculate the volume of material required to make it.

9) A bucket used on a crane is in the form of a frustum of a pyramid. Its base is a square of 600 mm side and its top is a square of 750 mm side. It has a depth of 800 mm.
   a) Calculate, in cubic metres, the volume of cement that it will hold.
   b) Twenty of these buckets of cement are emptied into a cylindrical cavity 4 m diameter. Calculate the depth to which the cavity will be filled.

10) The ball of a float valve is a sphere 200 mm diameter. It is immersed in water to a depth of 100 mm. How many litres of water does it displace?

11) A hopper is in the form of a frustum of a rectangular pyramid. The top is 4 m long and 3 m wide, whilst the bottom is 2 m long and 1.5 m wide. If it is 2 m deep, calculate the volume of the hopper and the total surface area of the inclined faces.

12) A tub holding 58 litres of water when full is shaped like the frustum of a cone. If the radii at the ends of the tub are 400 mm and 300 mm respectively, caculate the height of the tub.

2

Algebraic expressions

Indices are an extremely useful tool in mathematics. One typical case is their use when large numbers, such as ten million—which is cumbersome to write—may be expressed concisely as 10^7. This is a neat form in which 'base' ten is written with an index seven. Indices, which may be negative or fractional, are also used in mathematical relationships such as the law of expansion of gases, or the decay of an electric current.

Logarithms—another tool in mathematics—are another way of expressing relationships which use indices. These are much easier to handle now that we have calculating machines; older people will tell nightmarish stories of their encounters with logarithmic tables, which are now virtually obsolete.

Indices

You may already be familiar with the basic idea of indices. However, we start from scratch and show that it is just as easy to use indices for letters in algebra as it is for numbers in arithmetic.

Index, base and power

Numbers may be written in many forms: you are probably familiar with 3^2. This is an example of a number with an index. The figure 2 gives the number of 3s to be multiplied together and is called the index (plural indices).

Thus $$3 \times 3 = 3^2$$

We also call 3 the base, and 3^2 is known as the second power of the base 3. Now we often use letters to represent numbers (algebra), and so the laws of arithmetic apply strictly to algebraic terms as well as numbers.

So, using letters, $$a \times a \times a = a^3$$

Here a^3 is the third power of the base a, and the index is 3.

Or $$y \times y \times y \times y = y^4$$

Here y^4 is the fourth power of the base y, and the index is 4.

Thus in the expression x^n	x^n	**is called the nth power of x**
	x	**is called the base**
	n	**is called the index**

The six laws of indices

Law 1: Multiplication of Powers

Let us see what happens when we multiply powers of the same base together.

Now
$$5^2 \times 5^4 = (5 \times 5) \times (5 \times 5 \times 5 \times 5)$$
$$= 5 \times 5 \times 5 \times 5 \times 5 \times 5 = 5^6$$

or
$$c^3 \times c^5 = (c \times c \times c) \times (c \times c \times c \times c \times c)$$
$$= c \times c \times c \times c \times c \times c \times c \times c = c^8$$

In both the examples above we see that we could have obtained the result by adding the indices together.

Thus
$$5^2 \times 5^4 = 5^{2+4} = 5^6$$
and
$$c^3 \times c^5 = c^{3+5} = c^8$$

In general terms the law is

$$a^m \times a^n = a^{m+n}$$

We may apply this idea when multiplying more than two powers of the same base together.

Thus
$$7^2 \times 7^5 \times 7^9 = 7^{2+5} \times 7^9 = 7^7 \times 7^9 = 7^{7+9} = 7^{16}$$

We see that the same result would have been obtained by adding the indices, hence

$$7^2 \times 7^5 \times 7^9 = 7^{2+5+9} = 7^{16}$$

The law is:

> When multiplying powers *of the same base* together, add the indices.

EXAMPLE 2.1

Simplify $m^5 \times m^4 \times m^6 \times m^2$

$$m^5 \times m^4 \times m^6 \times m^2 = m^{5+4+6+2} = m^{17}$$

Law 2: Division of Powers

Now let us see what happens when we divide powers of the same base

$$\frac{3^5}{3^2} = \frac{3 \times 3 \times 3 \times 3 \times 3}{3 \times 3} = 3 \times 3 \times 3 = 3^3$$

We see that the same result could have been obtained by subtracting the indices.

Thus
$$\frac{3^5}{3^2} = 3^{5-2} = 3^3$$

In general terms the law is

$$\frac{a^m}{a^n} = a^{m-n}$$

or

> When dividing powers *of the same base* subtract the index of the denominator from the index of the numerator.

EXAMPLE 2.2

a) $\dfrac{4^8}{4^5} = 4^{8-5} = 4^3$

b) $\dfrac{z^3 \times z^4 \times z^8}{z^5 \times z^6} = \dfrac{z^{3+4+8}}{z^{5+6}} = \dfrac{z^{15}}{z^{11}} = z^{15-11} = z^4$

Law 3: Power of Power

How do we simplify $(5^3)^2$?   One way is to proceed as follows:

$$(5^3)^2 = 5^3 \times 5^3 = 5^{3+3} = 5^6$$

We see that the same result would have been obtained if we multiplied the two indices together.

Thus
$$(5^3)^2 = 5^{3\times2} = 5^6$$

In general terms the law is

$$(a^m)^n = a^{mn}$$

or

When raising the power of a base to a power, multiply the indices together.

EXAMPLE 2.3

a) $(8^4)^3 = 8^{4 \times 3} = 8^{12}$

b) $(p^2 \times q^4)^3 = (p^2)^3 \times (q^4)^3 = p^{2\times3} \times q^{4\times3} = p^6 \times q^{12}$

c) $\left(\dfrac{a^7}{b^5}\right)^6 = \dfrac{(a^7)^6}{(b^5)^6} = \dfrac{a^{7\times6}}{b^{5\times6}} = \dfrac{a^{42}}{b^{30}}$

Law 4: Zero Index

Now
$$\frac{2^5}{2^5} = \frac{2 \times 2 \times 2 \times 2 \times 2}{2 \times 2 \times 2 \times 2 \times 2} = 1$$

but using the laws of indices

$$\frac{2^5}{2^5} = 2^{5-5} = 2^0$$

Thus
$$2^0 = 1$$

Also
$$\frac{c^4}{c^4} = \frac{c \times c \times c \times c}{c \times c \times c \times c} = 1$$

But using the laws of indices

$$\frac{c^4}{c^4} = c^{4-4} = c^0$$

Thus
$$c^0 = 1$$

In general terms the law is

$$x^0 = 1$$

or

Any base raised to the index of zero is equal to 1.

EXAMPLE 2.4

a) $25^0 = 1$ b) $(0.56)^0 = 1$ c) $\left(\tfrac{1}{4}\right)^0 = 1$

Law 5:   Negative Indices

Now $\dfrac{2^3}{2^7} = \dfrac{2 \times 2 \times 2}{2 \times 2 \times 2 \times 2 \times 2 \times 2 \times 2} = \dfrac{1}{2 \times 2 \times 2 \times 2} = \dfrac{1}{2^4}$

but using the laws of indices

$$\frac{2^3}{2^7} = 2^{3-7} = 2^{-4}$$

It follows that

$$2^{-4} = \frac{1}{2^4}$$

Also

$$\frac{d}{d^2} = \frac{d}{d \times d} = \frac{1}{d}$$

but using the laws of indices

$$\frac{d}{d^2} = \frac{d^1}{d^2} = d^{1-2} = d^{-1}$$

It follows that

$$d^{-1} = \frac{1}{d}$$

In general terms the law is

$$x^{-n} = \frac{1}{x^n}$$

EXAMPLE 2.5

a) $3^{-1} = \dfrac{1}{3^1} = \dfrac{1}{3}$

b) $5x^{-3} = 5 \times x^{-3} = 5 \times \dfrac{1}{x^3} = \dfrac{5}{x^3}$

c) $(2a)^{-4} = \dfrac{1}{(2a)^4} = \dfrac{1}{2^4 \times a^4} = \dfrac{1}{16a^4}$

d) $\dfrac{1}{z^{-5}} = \dfrac{1}{\dfrac{1}{z^5}} = 1 \div \dfrac{1}{z^5} = 1 \times \dfrac{z^5}{1} = z^5$

Summary of the meaning of positive, zero and negative indices:

$$x^4 = x \times x \times x \times x$$
$$x^3 = x \times x \times x$$
$$x^2 = x \times x$$
$$x^1 = x$$
$$x^0 = 1$$
$$x^{-1} = \frac{1}{x}$$
$$x^{-2} = \frac{1}{x \times x} = \frac{1}{x^2}$$
$$x^{-3} = \frac{1}{x \times x \times x} = \frac{1}{x^3}$$

Each line of the above sequence is obtained by dividing the previous line by x.

The above sequence may help you to appreciate the meaning of positive and negative indices, and especially the zero index.

Remember: $\qquad$ (elephants)$^0 = 1$

Exercise 2.1

Simpify the following, giving each answer as a power:

1) $2^5 \times 2^6$

2) $a \times a^2 \times a^5$

3) $n^8 \div n^5$

4) $3^4 \times 3^7$

5) $b^2 \div b^5$

6) $10^5 \times 10^3 \div 10^4$

7) $z^4 \times z^2 \times z^{-3}$

8) $3^2 \times 3^{-3} \div 3^3$

9) $\dfrac{m^5 \times m^6}{m^4 \times m^3}$

10) $\dfrac{x^2 \times x}{x^6}$

11) $(9^3)^4$

12) $(y^2)^{-3}$

13) $(t \times t^3)^2$

14) $(c^{-7})^{-2}$

15) $\left(\dfrac{a^2}{a^5}\right)^3$

16) $\left(\dfrac{1}{7^3}\right)^4$

17) $\left(\dfrac{b^2}{b^7}\right)^{-2}$

18) $\dfrac{1}{(s^3)^3}$

Without using tables or calculating machines find the values of the following:

19) $\dfrac{8^3 \times 8^2}{8^4}$

20) $\dfrac{7^2 \times 7^5}{7^3 \times 7^4}$

21) $\dfrac{2^2}{2^2 \times 2}$

22) $2^4 \times 2^{-1}$

23) 2^{-2}

24) $\dfrac{1}{(10)^{-2}}$

25) $\dfrac{2^{-1}}{2}$

26) $\dfrac{24^0}{7}$

27) $(5^{-1})^2$

28) $3^{-3} \div 3^{-4}$

29) $\dfrac{7}{24^0}$

30) $\left(\dfrac{1}{5}\right)^{-2}$

31) 7×24^0

32) $\left(\dfrac{2}{3}\right)^{-3}$

33) $\left(\dfrac{2}{2^{-3}}\right)^{-2}$

Law 6: Fractional Indices

The cube root of 5 (written as $\sqrt[3]{5}$) is the number which, when multiplied by itself three times, gives 5.

Thus $$\sqrt[3]{5} \times \sqrt[3]{5} \times \sqrt[3]{5} = 5$$

But we also know that

$$5^{1/3} \times 5^{1/3} \times 5^{1/3} = 5^{1/3+1/3+1/3} = 5$$

Comparing these expressions

$$\sqrt[3]{5} = 5^{1/3}$$

Similarly the fourth root of base d written as ($\sqrt[4]{d}$) is the number which, when multiplied by itself four times, gives d.

Thus $$\sqrt[4]{d} \times \sqrt[4]{d} \times \sqrt[4]{d} \times \sqrt[4]{d} = d$$

But we also know that

$$d^{1/4} \times d^{1/4} \times d^{1/4} \times d^{1/4} = d^{1/4+1/4+1/4+1/4} = d$$

Comparing these expressions

$$\sqrt[4]{d} = d^{1/4}$$

In general terms the law is

$$\sqrt[n]{x} = x^{1/n}$$

Thus a fractional index represents a root—the denominator of the index denotes the root to be taken.

Summary of the six laws of indices:

Law 1	$a^m \times a^n = a^{m+n}$	Multiplication of powers
Law 2	$\dfrac{a^m}{a^n} = a^{m-n}$	Division of powers
Law 3	$(a^m)^n = a^{m \times n}$	Power of power
Law 4	$a^0 = 1$	Zero index
Law 5	$a^{-n} = \dfrac{1}{a^n}$	Negative index
Law 6	$a^{1/n} = \sqrt[n]{a}$	Fractional index

As you work through the text examples make sure that you can identify, at each step, which law is being used. In this way you will soon become familiar with each law and its use.

Number examples using indices

Here are some examples in which law 6 is used. We do appreciate that the numerical answers may be found from your calculator, but if you work through the solutions as they are given here, it should help your understanding of indices.

EXAMPLE 2.6

Find the value of a) $25^{1/2}$ and   b) $81^{1/4}$

a) $25^{1/2} = \sqrt{25} = 5$ (Note that for square roots the figure 2 indicating the root is usually omitted)

b) $81^{1/4} = \sqrt[4]{81} = 3$

EXAMPLE 2.7

Find the value of $8^{2/3}$

Now we may write $\frac{2}{3}$ as either (i) $2 \times \frac{1}{3}$ or (ii) $\frac{1}{3} \times 2$

So using (i) $8^{2/3} = 8^{2 \times \frac{1}{3}}$

$$= (8^2)^{\frac{1}{3}} \qquad \text{using law 3 for power of power}$$

$$= \sqrt[3]{8^2} \qquad \text{using law 6 for fractional indices}$$

$$= \sqrt[3]{64}$$

$$= 4$$

Alternatively

using (ii) $\qquad 8^{2/3} = 8^{\frac{1}{3} \times 2}$

$$= (8^{\frac{1}{3}})^2 \qquad \text{using law 3 for power of power}$$

$$= (\sqrt[3]{8})^2 \qquad \text{using law 6 for fractional indices}$$

$$= (2)^2$$

$$= 4$$

Both approaches give the same answer—but in this particular case which do you favour, and why?

EXAMPLE 2.8

Find the value of $16^{-3/4}$

In the previous example the second sequence is better as the arithmetic is simpler! This is even more important here, so we will use the fact that $\frac{3}{4} = \frac{1}{4} \times 3$

So $\quad 16^{-3/4} = \dfrac{1}{16^{3/4}}$ using law 5 for negative indices

$\qquad\qquad = \dfrac{1}{16^{\frac{1}{4}\times 3}}$

$\qquad\qquad = \dfrac{1}{(16^{\frac{1}{4}})^3}$ using law 3 for power of power

$\qquad\qquad = \dfrac{1}{(\sqrt[4]{16})^3}$ using law 6 for fractional indices

$\qquad\qquad = \dfrac{1}{(2)^3}$

$\qquad\qquad = \dfrac{1}{8}$

$\qquad\qquad = 0.125$

EXAMPLE 2.9

Find the value of $\quad \dfrac{1}{(\sqrt{7})^{-2}}$

$\dfrac{1}{(\sqrt{7})^{-2}} = \dfrac{1}{\dfrac{1}{(\sqrt{7})^2}}$ using law 5 for negative indices

$\qquad\qquad = 1 \div \dfrac{1}{(\sqrt{7})^2}$

$\qquad\qquad = 1 \times \dfrac{(\sqrt{7})^2}{1}$

$\qquad\qquad = (\sqrt{7})^2$

$\qquad\qquad = (7^{\frac{1}{2}})^2$ using law 6 for fractional indices

$\qquad\qquad = 7^{\frac{1}{2}\times 2}$ using law 3 for power of power

$\qquad\qquad = 7^1$

$\qquad\qquad = 7$

Exercise 2.2

Without using tables or calculating machines find the values of the following:

1) $5^2 \times 5^{1/2} \times 5^{-3/2}$ 　　　2) $4 \div 4^{1/2}$ 　　　3) $8^{1/3}$

4) $64^{1/6}$ 　　　5) $8^{2/3}$ 　　　6) $25^{3/2}$

7) $(16^{1/4})^3$ 　　　8) $\dfrac{1}{9^{-3/2}}$ 　　　9) $\left(\dfrac{1}{4}\right)^{-1/2}$

10) $16^{0.5}$ 　　　11) $36^{-0.5}$ 　　　12) $(4^{-3})^{1/2}$

13) $\left(\dfrac{1}{4}\right)^{5/2}$ 　　　14) $\left(\dfrac{1}{16^{0.5}}\right)^{-3}$ 　　　15) $\dfrac{1}{(\sqrt{3})^{-2}}$

Algebra examples using indices

EXAMPLE 2.10
Express each of the following as a single power:

a) $\sqrt{x^3}$ b) $\dfrac{1}{\sqrt[4]{a^5}}$ c) $\dfrac{(\sqrt{x})^4 x^{\frac{3}{4}}}{x^{\frac{1}{4}}}$

a)

$$\sqrt{x^3} = (x^3)^{\frac{1}{2}} \qquad \text{using law 6 for fractional indices}$$

$$= x^{3 \times \frac{1}{2}} \qquad \text{using law 3 for power of power}$$

$$= x^{3/2} = x^{1.5}$$

b)

$$\frac{1}{\sqrt[4]{a^5}} = \frac{1}{(a^5)^{\frac{1}{4}}} \qquad \text{using law 6 for fractional indices}$$

$$= \frac{1}{a^{5 \times \frac{1}{4}}} \qquad \text{using law 3 for power of power}$$

$$= \frac{1}{a^{5/4}}$$

$$= a^{-5/4} \qquad \text{using law 5 for negative index}$$

$$= a^{-1.25}$$

c)

$$\frac{(\sqrt{x})^4 x^{\frac{3}{4}}}{x^{\frac{1}{4}}} = \frac{(x^{\frac{1}{2}})^4 x^{\frac{3}{4}}}{x^{\frac{1}{4}}} \qquad \text{using law 6 for fractional indices}$$

$$= \frac{x^{\frac{1}{2} \times 4} x^{\frac{3}{4}}}{x^{\frac{1}{4}}} \qquad \text{using law 3 for power of power}$$

$$= \frac{x^2 x^{\frac{3}{4}}}{x^{\frac{1}{4}}} = \frac{x^{2+\frac{3}{4}}}{x^{\frac{1}{4}}} \qquad \text{using law 1 for multiplication of powers}$$

$$= \frac{x^{\frac{11}{4}}}{x^{\frac{1}{4}}} = x^{\frac{11}{4}-\frac{1}{4}} \qquad \text{using law 2 for division of powers}$$

$$= x^{\frac{10}{4}} \quad \text{or} \quad x^{2.5}$$

Exercise 2.3

Simplify the following, expressing each answer as a single power:

1) $\sqrt{x}$ 2) $\sqrt[5]{x^4}$ 3) $\dfrac{1}{\sqrt{x}}$

4) $\dfrac{1}{\sqrt[3]{1x}}$ 5) $\dfrac{1}{\sqrt[3]{x^4}}$ 6) $\sqrt{x^{-3}}$

7) $\dfrac{1}{\sqrt[3]{x^{-2}}}$ 8) $\dfrac{1}{\sqrt[4]{x^{-0.3}}}$ 9) $(\sqrt[3]{-x})^2$

10) $\sqrt{x^{2/3}}$ 11) $(\sqrt{x})^{2/3}$ 12) $\left(\dfrac{1}{\sqrt[3]{x^4}}\right)^{-3/4}$

13) $\dfrac{\sqrt[3]{a}}{a^2 \times \sqrt{a}}$ 14) $\dfrac{a^{-3}}{a^{2/3}}$ 15) $\dfrac{x^3}{\sqrt{x^{-1.5}}}$

16) $\dfrac{b^{5/2} \times b^{-3/2}}{b^{1/2}}$

17) $\dfrac{m^{-3/4}}{m^{-5/2}}$

18) $\dfrac{z^{2.3} \times z^{-1.5}}{z^{-3.5} \times z^2}$

19) $\dfrac{(x^{1/2})^3}{(x^3)^{1/2}}$

20) $\dfrac{\sqrt{u}}{u^3}$

21) $\dfrac{\sqrt[4]{y^3}}{\sqrt{y}}$

22) $\dfrac{(\sqrt[4]{n})^3}{\sqrt{n}}$

23) $\dfrac{\sqrt[4]{x^2}}{\sqrt[7]{x^{-2}}}$

24) $\dfrac{\sqrt[3]{t} \times \sqrt{t^3}}{t^{5/2}}$

Logarithms and exponent calculations

We will first see how the logarithmic definition is related to base numbers and indices, and then move on to the laws of logarithms so that we can make the use of logarithms as versatile as possible. We then use logarithms for numerical calculations and extend their use to the solving of indicial equations.

Logarithms

We are familiar with an expression such as $\qquad 9 = 3^2$
A similar algebraic equation is $\qquad N = b^x$

Now suppose we wish to find the value of x. It will be necessary to transpose the equation and make x the subject. Since we cannot do this using the usual methods we have an alternative form using logarithms (logs).

> So if $N = b^x$ our alternative form is $x = \log_b N$
>
> which in words is 'x is the logarithm of N to the base b'

> It is sometimes written the other way round as $\log_b N = x$
>
> which in words is 'the logarithm of N, to the base b, is x'

Some examples are:

We may write	$8 = 2^3$	We may write	$81 = 3^4$
in log form as	$\log_2 8 = 3$	in log form as	$\log_3 81 = 4$
We may write	$2 = \sqrt{4}$	We may write	$\dfrac{1}{4} = \dfrac{1}{2^2}$
or	$2 = 4^{1/2}$		
or	$2 = 4^{0.5}$	or	$0.25 = 2^{-2}$
in log form as	$\log_4 2 = 0.5$	in log form as	$\log_2 0.25 = -2$

EXAMPLE 2.11

If $\log_7 49 = x$, find the value of x.

Writing the equation in index form we have $\quad 49 = 7^x$

$$\text{or} \quad 7^2 = 7^x$$

Since the bases are the same on both sides of the equation the indices must be the same.

$$\text{Thus} \quad x = 2$$

EXAMPLE 2.12

If $\log_b 8 = 3$, find the value of b.

Writing this equation in index form we have $\quad 8 = b^3$

$$\text{or} \quad 2^3 = b^3$$

Since the indices on both sides of the equation are the same the bases must be the same.

$$\text{Thus} \quad b = 2$$

Some important values of logarithms

The value of $\log_b 1$

Let $\qquad\qquad\qquad \log_b 1 = x$

then in index form $\qquad\qquad 1 = b^x$

Now the only value of the index x which will satisfy this expression is zero.

Hence $\qquad\qquad\qquad \log_b 1 = 0$

> Thus: $\qquad$ To any base the value of log 1 is zero

The value of $\log_b b$

Let $\qquad\qquad\qquad \log_b b = x$

then in index form $\qquad\qquad b = b^x$

Now the only value of the index x which will satisfy the expression is unity.

Hence $\qquad\qquad\qquad \log_b b = 1$

> Thus: $\qquad$ The value of the log of a number to the same base is unity

The value of $\log_b(-N)$

Let $$\log_b(-N) = x$$
then in index form $$-N = b^x$$

If we examine this expression we can see that whatever the value of the negative number, N, or whatever the value of the base, b, it is not possible to find a value for the index x which will satisfy the expression.

Hence $\log_b(-N)$ has no real value

Thus: Only positive numbers have real logarithms

Exercise 2.4

Express in logarithmic form:

1) $n = a^x$
2) $2^3 = 8$
3) $5^{-2} = 0.04$
4) $10^{-3} = 0.001$
5) $x^0 = 1$
6) $10^1 = 10$
7) $a^1 = a$
8) $e^2 = 7.39$
9) $10^0 = 1$

Find the value of x in each of the following:

10) $\log_x 9 = 2$
11) $\log_x 81 = 4$
12) $\log_2 16 = x$
13) $\log_5 125 = x$
14) $\log_3 x = 2$
15) $\log_4 x = 3$
16) $\log_{10} x = 2$
17) $\log_7 x = 0$
18) $\log_x 8 = 3$
19) $\log_x 27 = 3$
20) $\log_9 3 = x$
21) $\log_n n = x$

Bases of logarithms

Although we can have logarithms to any base, in practice only two are used—they are base 10 and base e.

Logarithms to the base 10

Logarithms to the base 10 are called common logarithms and are stated as $\log_{10}$, log or lg. When logarithmic tables are used to solve numerical problems, tables to this base are preferred as they are simpler to use than tables to any other base. Common logarithms are also used for scales on logarithmic graph paper and for calculations on the measurement of sound.

Logarithms to the base e

In higher mathematics all logarithms are taken to the base e, where $e = 2.718\,28$ correct to 5 d.p. Logarithms to this base are often called natural logarithms. They are also called Naperian or hyperbolic logarithms.

Natural logarithms are stated as $\log_e$ (or ln).

The three laws of logarithms

Law 1

Let $\qquad \log_b M = x \qquad$ and $\qquad \log_b N = y$

or in index form $\qquad M = b^x \qquad$ and $\qquad N = b^y$

Now $\qquad MN = b^x \times b^y$

$\therefore \qquad MN = b^{x+y}$

or in log form $\qquad \log_b MN = x + y$

or since $\qquad x = \log_b M$ and $y = \log_b N$

$$\text{then} \qquad \log_b MN = \log_b M + \log_b N$$

In words Law 1 is:

The logarithm of two numbers multiplied together may be found by adding their individual logarithms.

The example below shows how the law is used with numbers.

EXAMPLE 2.13
Find the value of $\log_{10}(5 \times 6)$ using Law of 1 logarithms.

Using Law 1 then $\qquad \log_{10}(5 \times 6) = \log_{10} 5 + \log_{10} 6$

$\boxed{AC}\ \boxed{5}\ \boxed{\lg}\ \boxed{+}\ \boxed{6}\ \boxed{\lg}\ \boxed{=}$ displaying $\quad 1.477\,1213$

This shows that $\quad \log_{10}(5 \times 6) = 1.477 \quad$ correct to 3 d.p.

Alternatively using the simpler sequence

$\boxed{AC}\ \boxed{5}\ \boxed{\times}\ \boxed{6}\ \boxed{=}\ \boxed{\lg}$ displaying $\quad 1.477\,1213$

This shows that $\log_{10}(5 \times 6)$ and $\log_{10} 30$ give the same result and confirms the use of Law 1.

Law 2

Let $\qquad \log_b M = x \qquad$ and $\qquad \log_b N = y$

or in index form $\qquad M = b^x \qquad$ and $\qquad N = b^y$

Now
$$\frac{M}{N} = \frac{b^x}{b^y}$$
$$\frac{M}{N} = b^{x-y}$$

or in log form $\qquad \log_b \dfrac{M}{N} = x - y$

or since $\qquad x = \log_b M$ and $y = \log_b N$

then
$$\log_b \frac{M}{N} = \log_b M - \log_b N$$

The logarithm of two numbers divided may be found by subtracting their individual logarithms.

EXAMPLE 2.14

Find the value of $\log_e(\frac{6}{4})$ using Law 2 of logarithms.

Using Law 2 $\qquad \log_e(\frac{6}{4}) = \log_e 6 - \log_e 4$

then $\qquad$ $\boxed{AC}\,\boxed{6}\,\boxed{\ln}\,\boxed{-}\,\boxed{4}\,\boxed{\ln}\,\boxed{=}$ $\qquad$ displaying $\qquad$ 0.405 4651

This shows that $\log_e(\frac{6}{4}) = 0.405$ (correct to 3 d.p.)

You may care to check this result by using $\log_e(\frac{6}{4}) = \log_e 1.5$

Law 3

Let $\qquad \log_b M = x$

or in index form $\qquad M = b^x$

Now $\qquad M^n = (b^x)^n$

$\qquad M^n = b^{nx}$

or in log form $\qquad \log_b M^n = nx \qquad$ and since $x = \log_b M$

then $\qquad$ $\log_b M^n = n(\log_b M)$

The logarithm of a number raised to a power may be found by multiplying the logarithm of the number by the index.

EXAMPLE 2.15

Find the value of $\log_{10} 5^3$ using Law 3 of logarithms.

Using Law 3 $\qquad \log_{10} 5^3 = 3\log_{10} 5$

A suitable sequence is $\qquad$ $\boxed{5}\,\boxed{\lg}\,\boxed{\times}\,\boxed{3}\,\boxed{=}$ $\qquad$ displaying $\qquad$ 2.09691

This shows that $\log_{10} 5^3 = 2.097$ (correct to 3 d.p.)

You may care to check this using $\log_{10} 5^3 = \log_{10} 125$

Combining the laws

The algebraic example below uses all three laws.

EXAMPLE 2.16

Simplify $\log \dfrac{ab}{c^2}$ in terms of $\log a$, $\log b$ and $\log c$.

Base is of no importance here—our manipulation will suit any base.

$$
\begin{aligned}
\text{Now} \quad \log \frac{ab}{c^2} &= \log ab - \log c^2 && \text{using Law 2 of logs for} \\
&&& \text{numbers divided} \\
&= \log a + \log b - \log c^2 && \text{using Law 1 of logs for} \\
&&& \text{numbers multiplied} \\
&= \log a + \log b - 2\log c && \text{using Law 3 of logs for} \\
&&& \text{number powers}
\end{aligned}
$$

Calculations using logarithms

Most calculators have keys for both $\log_e$ (ln) and $\log_{10}$ (log or lg) so we may calculate using logarithms of either base.

To find the logarithm of a number

EXAMPLE 2.17
Find the value of a) $\log_e 3.4$ and b) $\log_{10} 0.876$

a) Use the sequence $\boxed{\text{AC}}\boxed{3.4}\boxed{\text{ln}}$ displaying 1.223 7754

This shows that $\log_e 3.4 = 1.224$ correct to 3 d.p.

b) Use the sequence $\boxed{\text{AC}}\boxed{0.876}\boxed{\text{lg}}$ displaying $-0.057\,4958$

This shows that $\log_{10} 0.876 = -0.0575$ correct to 4 d.p.

To find a number when given its logarithm

EXAMPLE 2.18
Find the number whose: a) natural logarithm is 0.5461
b) logarithm to the base 10 is 1.723

a) If N is our number then $\log_e N = 0.5461$

or, in index form, $N = e^{0.5461}$

This process is called finding the antilogarithm, to the base e, of 0.5461 and the calculator sequence is:

$\boxed{\text{AC}}\boxed{0.5461}\boxed{e^x}$ giving 1.726 5065

So our number is 1.7265 correct to 4 d.p.

or, in other words, antilog$_e$ 0.5461 = 1.7265

b) Here our calculator sequence will be to find $10^{1.723}$ and is:

$\boxed{\text{AC}}\boxed{1.723}\boxed{10^x}$ giving 52.844 525

Thus the number is 52.84 correct to 2 d.p.

or, in other words, antilog$_{10}$ 1.723 = 52.84

Summary of the laws of logarithms:

Law 1	$\log_b MN = \log_b M + \log bN$
Law 2	$\log_b \dfrac{M}{N} = \log_b M - \log_b N$
Law 3	$\log_b M^n = n(\log_b M)$

Exercise 2.5

1) Using the laws of logarithms express the following in terms of $\log a$, $\log b$, $\log c$ or $\log d$ as appropriate:

   a) $\log a^2 b$ b) $\log \dfrac{ac^3}{b^4}$ c) $\log \dfrac{ab}{cd}$

2) Find the values, correct to 3 d.p., of:

   a) $\log_e 3.76$ b) $\log_e 0.34$ c) $\log_{10} 35$ d) $\log_{10} 0.078$

3) Find the numbers, correct to 4 s.f., whose natural logarithms are:

   a) 2.76   b) 0.09   c) -3.46 d) -0.543

4) Find the numbers, correct to 3 s.f., whose logarithms to the base 10 are:

   a) 1.93   b) 0.297   c) -0.0056 d) -0.576

5) Evaluate $\log_e 12$. Check the value obtained by finding the value of $\log_e(3 \times 4)$, resulting from the use of logarithmic Law 1.

6) Evaluate $\log_{10} 1.25$ and check your result by using Law 2 of logarithms on $\log_{10}(\frac{5}{4})$ and finding the answer.

7) Evaluate $\ln 32$ and verify your result by finding the value of $\ln 2^5$ and using the third law of logarithms.

Indicial equations

These are equations in which the number to be found is an index, or part of an index.

The method of solution is to reduce the given equation to an equation involving logarithms, as the following examples will illustrate.

EXAMPLE 2.19
If $8.79^x = 67.95$ find the value of x.

Now taking logarithms of both sides of the given equation we have

$$\log 8.79^x = \log 67.65$$

$$\therefore \qquad x(\log 8.79) = \log 67.65$$

$$\therefore \qquad x = \frac{\log 67.65}{\log 8.79}$$

The base of the logarithms has not yet been chosen, the above procedure being true for any base value.

A good approach, having chosen which base to use, is to find the value of the bottom line and put this into the memory. Then find the value of the top line and divide this by the content of the memory.

Thus the answer is 1.94 correct to three significant figures.

It is good idea to check your answer by substituting 1.94 in the LHS of the original equation and verifying that it gives the RHS value—try this for yourself.

EXAMPLE 2.20
Find the value of x if $1.793^{(x+3)} = 20^{0.982}$

Now taking logarithms of both sides of the given equation we have

$$\log 1.793^{(x+3)} = \log 20^{0.982}$$

$$(x+3)(\log 1.793) = (0.982)(\log 20)$$

$$x + 3 = \frac{(0.982)(\log 20)}{\log 1.793}$$

$$x = \frac{(0.982)(\log 20)}{\log 1.793} - 3$$

$$= 2.04 \quad \text{correct to 3 s.f.}$$

Exercise 2.6

Evaluate the following:

1) $11.57^{0.3}$ 　　　　2) $15.62^{2.15}$ 　　　　3) $0.6327^{0.5}$

4) $0.065\ 21^{3.16}$ 　　　5) $27.15^{-0.4}$

Find the value of x in the following:

6) $3.6^x = 9.7$ 　　　　　　　7) $0.9^x = 2.176$

8) $\left(\dfrac{1}{7.2}\right)^x = 1.89$ 　　　　　9) $1.4^{(x+2)} = 9.3$

10) $21.9^{(3-x)} = 7.334$ 　　　　11) $2.79^{(x-1)} = 4.377^x$

12) $\left(\dfrac{1}{0.64}\right)^{(2+x)} = 1.543^{(x+1)}$ 　　　13) $\dfrac{1}{0.9^{(x-2)}} = 8.45$

Calculations involving exponent functions, e^x and e^{-x}

The exponent functions e^x and e^{-x} are of interest to engineers mainly because they are expressions which represent growth and decay—for instance, in electrical circuits. Later you will see their graphs but it is important, at this stage, that you should be able to handle expressions in which these functions occur.

EXAMPLE 2.21
Evaluate $50\,e^{2.16}$

The sequence of operations is:

$$\boxed{AC}\ \boxed{2}\ \boxed{.}\ \boxed{1}\ \boxed{6}\ \boxed{e^x}\ \boxed{\times}\ \boxed{5}\ \boxed{0}\ \boxed{=}$$

giving an answer 434 correct to 3 s.f.

EXAMPLE 2.22
Evaluate $200\,e^{-1.34}$

The sequence of operations would then be:

$$\boxed{AC}\ \boxed{1}\ \boxed{.}\ \boxed{3}\ \boxed{4}\ \boxed{+/-}\ \boxed{e^x}\ \boxed{\times}\ \boxed{2}\ \boxed{0}\ \boxed{0}\ \boxed{=}$$

giving an answer 52.4 correct to 3 s.f.

EXAMPLE 2.23

The formula
$$R = \frac{(0.42)S}{l} \times \log_e \frac{d_2}{d_1}$$

refers to the insulation resistance of a wire. Find the value of R when $S = 2000$, $l = 120$, $d_1 = 0.2$ and $d_2 = 0.3$

Substituting the given values gives

$$R = \frac{0.42 \times 2000}{120} \times \log_e \frac{0.3}{0.2}$$

$$= \frac{0.42 \times 2000}{120} \times \log_e 1.5$$

$$= 2.84 \quad \text{correct to 3 s.f.}$$

EXAMPLE 2.24

In a capacitive circuit the instantaneous voltage across the capacitor is given by $v = V(1 - e^{-t/CR})$ where V is the initial supply voltage, R ohms the resistance, C farads the capacitance and t seconds the time from the instant of connecting the supply voltage.

If $V = 200$, $R = 10\,000$ and $C = 20 \times 10^{-6}$ find the time when the voltage v is 100 volts.

Substituting the given values in the equation we have

$$100 = 200(1 - e^{t/20 \times 10^{-6} \times 10\,000})$$

$$\frac{100}{200} = 1 - e^{-t/0.2}$$

$$0.5 = 1 - e^{-5t}$$

$$e^{-5t} = 1 - 0.5$$

$$e^{-5t} = 0.5$$

Thus in log form

$$\log_e 0.5 = -5t$$

$$t = -\frac{\log_e 0.5}{5}$$

The sequence of operation is:

$$\boxed{\text{AC}}\,\boxed{0}\,\boxed{\cdot}\,\boxed{5}\,\boxed{\ln}\,\boxed{\div}\,\boxed{5}\,\boxed{=}\,\boxed{+/-}$$

giving an answer 0.139 seconds correct to 3 s.f.

Exercise 2.7

1) Find the values of:

   a) $70\,e^{2.5}$ b) $150\,e^{-1.34}$ c) $3.4\,e^{-0.445}$

2) The formula $L = 0.000\,644\left(\log_e \frac{d}{r} + \frac{1}{4}\right)$ is used for calculating the self-inductance of parallel conductors. Find L when $d = 50$ and $r = 0.25$.

3) The inductance (L microhenrys) of a straight aerial is given by the formula: $L = \frac{1}{500}\left(\log_e \frac{4l}{d} - 1\right)$ where l is the length of the aerial in mm and d its diameter in mm. Calculate the inductance of an aerial 5000 mm long and 2 mm in diameter.

4) Find the value of $\log_e\left(\frac{c_1}{c_2}\right)^2$ when $c_1 = 4.7$ and $c_2 = 3.5$

5) If $T = R\log_e\left(\frac{a}{a-b}\right)$ find T when $R = 28$, $a = 5$ and $b = 3$

6) When a chain of length $2l$ is suspended from two points $2d$ apart on the same horizontal level, $d = c \log_e \left(\dfrac{l + \sqrt{l^2 + c^2}}{c} \right)$. If $c = 80$ and $l = 200$ find d.

7) The instantaneous value of the current when an inductive circuit is discharging is given by the formula $i = I e^{-Rt/L}$. Find the value of this current, i, when $I = 6$, $R = 30$, $L = 0.5$ and $t = 0.005$

8) In a circuit in which a resistor is connected in series with a capacitor the instantaneous voltage across the capacitor is given by the formula $v = V(1 - e^{-t/CR})$. Find this voltage, v, when $V = 200$, $C = 40 \times 10^{-6}$, $R = 100\,000$ and $t = 1$.

9) In the formula $v = V e^{-Rt/L}$ the values of v, V, R and L are 50, 150, 60 and 0.3, respectively. Find the corresponding value of t.

10) The instantaneous charge in a capacitive circuit is given by $q = Q(1 - e^{-t/CR})$. Find the value of t when $q = 0.01$, $Q = 0.015$, $C = 0.0001$ and $R = 7000$.

3

Algebraic manipulation

This chapter covers the ways in which we approach and handle expressions arising from the application of algebra to engineering problems. Situations are often modelled as algebraic expressions and it is essential that these can be simplified into a form suitable for finding a solution.

Positive and negative numbers

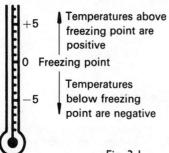

Temperatures above freezing point are positive

0 Freezing point

Temperatures below freezing point are negative

Fig. 3.1

We are all familiar with a Celsius thermometer; Fig. 3.1 shows part of one. The freezing point of water is 0°C (zero degrees Celsius). Temperatures above and below freezing may be read directly off the scale.

We now have to decide on a method for showing whether a temperature is above or below zero. We may say that a temperature is 5 degrees above zero or 6 degrees below zero, but these statements are not compact enough for calculations.

So we write them as +5°C and −6°C, and we have examples of a positive and a negative number.

When multiplying or dividing positive and negative numbers we use the rules:

> Two numbers with   like   signs give a positive result.
> Two numbers with   unlike   signs give a negative result.

As usual, we are lazy, and the + sign is usually omitted, so +12 is written simply as 12.

Some typical examples of the above rules are:

$7 \times 4 = 28$		$(-7) \times (4) = -28$	
$7 \times (-4) = -28$		$(-7) \times (-4) = 28$	
$\dfrac{20}{4} = 5$	$\dfrac{20}{-4} = -5$	$\dfrac{-20}{4} = -5$	$\dfrac{-20}{-4} = 5$

The modulus of a number

Sometimes we are only interested in the size (or magnitude) of a number irrespective of whether it is positive or negative. This magnitude is called the modulus, and if x is the number then it is written as

> $|x|$ and pronounced   'mod x'

Such an occasion is when using the theorem of Pythagoras to find a length which is given by a square root value, say for instance $\sqrt{4}$. We are only interested in the magnitude, 2, of the number, although strictly $\sqrt{4} = +2$ or -2.

Operations on algebraic quantities

Multiplication and division

The rules are exactly the same as those used with numbers:

> The product of two expressions with like signs is positive.

$$(+x)(+y) = +(xy) = +xy = xy$$
$$5x \times 3y = 5 \times 3 \times x \times y = 15xy$$
$$(-3a)(-2b) = +(3a)(2b) = 6ab$$

The product of two expressions with unlike signs is negative.

$$(-4x)(2y) = -(4x)(2y) = -8xy$$
$$(5p)(-6q) = -(5p)(6q) = -30pq$$

The result of dividing two expressions with like signs is positive.

$$\frac{+2c}{+d} = +\frac{2c}{d} = \frac{2c}{d} \qquad \frac{-3m}{-9n} = +\frac{3m}{9n} = \frac{m}{3n}$$

The result of dividing two expressions with unlike signs is negative.

$$\frac{-5x}{2y} = -\frac{5x}{2y} \qquad\qquad \frac{4r}{-s} = -\frac{4r}{s}$$

When *multiplying* expressions containing the same symbols, indices are used:

$$m \times m = m^2$$
$$3m \times 5m = 3 \times m \times 5 \times m = 15m^2$$
$$(-m) \times m^2 = (-m) \times m \times m = -m^3$$
$$5m^2n \times 3mn^3 = 5 \times m \times m \times n \times 3 \times m \times n \times n \times n = 15m^3n^4$$
$$3mn \times (-2n^2) = 3 \times m \times n \times (-2) \times n \times n = -6mn^3$$

When *dividing* algebraic expressions, cancellation between numerator and denominator is often possible. Cancelling is equivalent to dividing both numerator and denominator by the same quantity:

Thus
$$\frac{pq}{p} = \frac{\cancel{p} \times q}{\cancel{p}} = q$$

or
$$\frac{3p^2q}{6pq^2} = \frac{3 \times \cancel{p} \times p \times \cancel{q}}{6 \times \cancel{p} \times \cancel{q} \times q} = \frac{3p}{6q} = \frac{p}{2q}$$

or
$$\frac{18x^2y^2z}{6xyz} = \frac{18 \times \cancel{x} \times x \times \cancel{y} \times y \times \cancel{z}}{6 \times \cancel{x} \times \cancel{y} \times \cancel{z}} = 3xy$$

Sequence of mixed operations

Algebraic quantities contain symbols (or letters) which represent numbers. Thus the sequence of operations, namely:

> Brackets, Of, Divide, Multiply, Add, Subtract,

when simplifying algebraic expressions is exactly the same as used for number expressions.

Remember the mnemonic 'BODMAS' which gives the initial letters in the correct order.

Thus
$$\begin{aligned} 2x^2 + (12x^4 - 3x^4) \div 3x^2 - x^2 &= 2x^2 + 9x^4 \div 3x^2 - x^2 \\ &= 2x^2 + 3x^2 - x^2 \\ &= 5x^2 - x^2 \\ &= 4x^2 \end{aligned}$$

Brackets

Brackets are used for convenience in grouping terms together. When removing brackets each *term* within the bracket is multiplied by the quantity outside the bracket:

$$3(x + y) = 3x + 3y$$
$$5(2x + 3y) = 5 \times 2x + 5 \times 3y = 10x + 15y$$
$$4(a - 2b) = 4 \times a - 4 \times 2b = 4a - 8b$$
$$m(a + b) = ma + mb$$
$$3x(2p + 3q) = 3x \times 2p + 3x \times 3q = 6px + 9qx$$
$$4a(2a + b) = 4a \times 2a + 4a \times b = 8a^2 + 4ab$$

When a bracket has a minus sign in front of it, the signs of all the terms inside the bracket are changed when the bracket is removed. The reason for this rule may be seen from the following examples:

$$-3(2x - 5y) = (-3) \times 2x + (-3) \times (-5y) = -6x + 15y$$
$$-(m + n) = -m - n$$
$$-(p - q) = -p + q$$
$$-2(p + 3q) = -2p - 6q$$

When simplifying expressions containing brackets first remove the brackets and then add the like terms together:

$$(3x + 7y) - (4x + 3y) = 3x + 7y - 4x - 3y = -x + 4y$$
$$3(2x + 3y) - (x + 5y) = 6x + 9y - x - 5y = 5x + 4y$$
$$x(a + b) - x(a + 3b) = ax + bx - ax - 3bx = -2bx$$
$$2(5a + 3b) + 3(a - 2b) = 10a + 6b + 3a - 6b = 13a$$

Exercise 3.1

1) $3(3x + 2y)$

2) $5(2p - 3q)$

3) $-(a - 2b)$

4) $-4(x + 3)$

5) $2k(k - 5)$

6) $-3y(3x + 4)$

7) $a(p - q - r)$

8) $4xy(ab - ac + d)$

9) $3x^2(x^2 - 2xy + y^2)$

10) $-7P(2P^2 - P + 1)$

11) $-2m(-1 + 3m - 2n)$

Remove the brackets and simplify the following:

12) $3(x + 4) - (2x + 5)$

13) $4(1 - 2x) - 3(3x - 4)$

14) $5(2x - y) - 3(x + 2y)$

15) $\frac{1}{2}(y - 1) + \frac{1}{3}(2y - 3)$

16) $-(4a + 5b - 3c) - 2(2a + 3b - 4c)$

17) $3(a - b) - 2(2a - 3b) + 4(a - 3b)$

18) $3x(x^2 + 7x - 1) - 2x(2x^2 + 3) - 3(x^2 + 5)$

Binominal expressions

A binomial expression consists of two terms.
Thus $3x + 5$, $a + b$, $2x + 37$ and $4p - q$ are all binomial expressions.

The product of two binomial expressions

To find the product of $(a + b)(c + d)$ consider the diagram (Fig. 3.2).

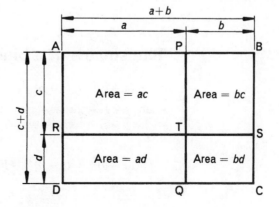

Fig. 3.2

In Fig. 3.2 the rectangular area ABCD is made up as follows:

$$ABCD = APTR + TQDR + PBST + TSCQ$$

i.e. $(a + b)(c + d) = ac + ad + bc + bd$

It will be noticed that the expression on the right hand side is obtained by multiplying each term in the one bracket by each term in the other bracket. The process is illustrated below:

$$(a + b)(c + d) = a(c + d) + b(c + d)$$
$$= ac + ad + bc + bd$$

EXAMPLE 3.1

a) $(3x + 2)(4x + 5) = 3x \times 4x + 3x \times 5 + 2 \times 4x + 2 \times 5$
$$= 12x^2 + 15x + 8x + 10$$
$$= 12x^2 + 23x + 10$$

b) $(2p - 3)(4p + 7) = 2p \times 4p + 2p \times 7 - 3 \times 4p - 3 \times 7$
$$= 8p^2 + 14p - 12p - 21$$
$$= 8p^2 + 2p - 21$$

c) $(z - 5)(3z - 2) = z \times 3z + z \times (-2) - 5 \times 3z - 5 \times (-2)$
$$= 3z^2 - 2z - 15z + 10$$
$$= 3z^2 - 17z + 10$$

d) $(2x + 3y)(3x - 2y) = 2x \times 3x + 2x \times (-2y) + 3y \times 3x$
$$+ 3y \times (-2y)$$
$$= 6x^2 - 4xy + 9xy - 6y^2$$
$$= 6x^2 + 5xy - 6y^2$$

The square of a binomial expression

$$(a + b)^2 = (a + b)(a + b) = a^2 + ab + ab + b^2 = a^2 + 2ab + b^2$$

$\therefore$ $(a + b)^2 = a^2 + 2ab + b^2$

$$(a - b)^2 = (a - b)(a - b) = a^2 - ab - ba + b^2 = a^2 - 2ab + b^2$$

$\therefore$ $(a - b)^2 = a^2 - 2ab + b^2$

EXAMPLE 3.2

a)
$$(2x + 5)^2 = (2x)^2 + 2 \times 2x \times 5 + 5^2$$
$$= 4x^2 + 20x + 25$$

b)
$$(3x - 2)^2 = (3x)^2 + 2 \times 3x \times (-2) + (-2)^2$$
$$= 9x^2 - 12x + 4$$

c)
$$(2x + 3y)^2 = (2x)^2 + 2 \times 2x \times 3y + (3y)^2$$
$$= 4x^2 + 12xy + 9y^2$$

The product of the sum and difference of two terms

$$(a + b)(a - b) = a^2 - ab + ba - b^2$$

$$(a + b)(a - b) = a^2 - b^2$$

This result is the difference of the squares of the two terms.

EXAMPLE 3.3

a)
$$(8x + 3)(8x - 3) = (8x)^2 - 3^2 = 64x^2 - 9$$

b)
$$(2x + 5y)(2x - 5y) = (2x)^2 - (5y)^2 = 4x^2 - 25y^2$$

Simplifying three brackets

If you are happy multiplying out two brackets, then this will seem easy!

Consider $\qquad (x - 2)(x + 3)(2x + 7)$

First we simplify two of the brackets. So choosing the last two and multiplying out

$$(x + 3)(2x + 7) = x(2x + 7) + 3(2x + 7)$$
$$= x \times 2x + x \times 7 + 3 \times 2x + 3 \times 7$$
$$= 2x^2 + 7x + 6x + 21$$
$$= 2x^2 + 13x + 21$$

The problem is now reduced to

$$(x - 2)(2x^2 + 13x + 21) \quad \text{and if we multiply out as before}$$

then $\quad (x - 2)(2x^2 + 13x + 21) = x(2x^2 + 13x + 21) - 2(2x^2 + 13x + 21)$
$$= x \times 2x^2 + x \times 13x + x \times 21$$
$$\quad -2 \times 2x^2 - 2 \times 13x - 2 \times 21$$
$$= 2x^3 + 13x^2 + 21x - 4x^2 - 26x - 42$$
$$= 2x^3 + 9x^2 - 5x - 42$$

Exercise 3.2

Find the products of the following:

1) $(x+4)(x+5)$ 2) $(2x+5)(x+3)$

3) $(2x+4)(3x+2)$ 4) $(5x+1)(2x+3)$

5) $(7x+2)(3x+5)$ 6) $(x-1)(x-3)$

7) $(x+3)(x-1)$ 8) $(x-2)(x+7)$

9) $(x-5)(x+3)$ 10) $(2x+5)(x-2)$

11) $(3x+5)(2x-3)$ 12) $(6x-7)(2x+3)$

13) $(2p-q)(p-3q)$ 14) $(3v+2u)(2v-3u)$

15) $(2a+b)(3a-b)$ 16) $(x+1)^2$

17) $(2x+3)^2$ 18) $(3x+7)^2$

19) $(x-1)^2$ 20) $(2x-3)^2$

21) $(x+y)^2$ 22) $(P+3Q)^2$

23) $(3x-4y)^2$ 24) $(2x+y)(2x-y)$

25) $(2m-3n)(2m+3n)$ 26) $(x^2+y)(x^2-y)$

27) $(x+1)(x-2)(x+3)$ 28) $(2x-1)(x+2)(x-3)$

29) $(x+1)^2(x-2)$ 30) $(x+1)(x-1)(x^2+1)$

31) $(x+a)(x+b)(x+c)$

Highest common factor (HCF)

The HCF of a set of algebraic expressions is the highest expression which is a factor of each of the given expressions.

The method used is similar to that for finding the HCF of a set of numbers.

EXAMPLE 3.4
Find the HCF of ab^2c^2, $a^2b^3c^3$, $a^2b^4c^4$.

We express each expression as the product of its factors.

Thus $ab^2c^2 = a \times b \times b \times c \times c$

and $a^2b^3c^3 = a \times a \times b \times b \times b \times c \times c \times c$

and $a^2b^4c^4 = a \times a \times b \times b \times b \times b \times c \times c \times c \times c$

We now note the factors which are common to each of the lines. Factor a is common once, factor b twice, and factor c twice. The product of these factors gives the required HCF.

Thus   HCF $= a \times b \times b \times c \times c$

$= ab^2c^2$

EXAMPLE 3.5

Find the HCF of $\dfrac{x^3 y}{m^2 n^4}, \dfrac{x^2 y^3}{m^2 n^2}, \dfrac{x^4 y^2}{mn^3}$.

Now $\qquad \dfrac{x^3 y}{m^2 n^4} = x \times x \times x \times y \times \dfrac{1}{m} \times \dfrac{1}{m} \times \dfrac{1}{n} \times \dfrac{1}{n} \times \dfrac{1}{n} \times \dfrac{1}{n}$

and $\qquad \dfrac{x^2 y^3}{m^2 n^2} = x \times x \times y \times y \times y \times \dfrac{1}{m} \times \dfrac{1}{m} \times \dfrac{1}{n} \times \dfrac{1}{n}$

and $\qquad \dfrac{x^4 y^2}{mn^3} = x \times x \times x \times x \times y \times y \times \dfrac{1}{m} \times \dfrac{1}{n} \times \dfrac{1}{n} \times \dfrac{1}{n}$

Factor x is common twice, factor y once, factor $\dfrac{1}{m}$ once, and factor $\dfrac{1}{n}$ twice.

Thus $\qquad\qquad\qquad \text{HCF} = x \times x \times y \times \dfrac{1}{m} \times \dfrac{1}{n} \times \dfrac{1}{n}$

$$= \dfrac{x^2 y}{mn^2}$$

An alternative method is to select the lowest power of each of the quantities which occur in *all* of the expressions, and then multiply them together.

EXAMPLE 3.6

Find the HCF of $3m^2 np^3, \ 6m^3 n^2 p^2, \ 24m^3 p^4$.

Dealing with the numerical coefficients 3, 6 and 24 we note that 3 is a factor of each of them. The quantities m and p occur in all three expressions, their lowest powers being m^2 and p^2. Hence,

$$\text{HCF} = 3m^2 p^2$$

(Note that n does not occur in each of the three expressions and hence it does not appear in the HCF.)

Factorising

A factor is a common part of two or more terms which make up an algebraic expression. Thus the expression $3x + 3y$ has two terms which have the number 3 common to both of them. Thus $3x + 3y = 3(x + y)$. We say that 3 and $(x + y)$ are the factors of $3x + 3y$. To factorise algebraic expressions of this kind, we first find the HCF of all the terms making up the expression. The HCF then appears outside the bracket. To find the terms inside the bracket divide each of the terms making up the expression by the HCF.

EXAMPLE 3.7

a) Find the factors of $\ ax + bx$.

The HCF of ax and bx is x.

$\therefore \qquad\qquad ax + bx = x(a + b) \qquad$ since $\quad \dfrac{ax}{x} = a \ $ and $\ \dfrac{bx}{x} = b$

b) Find the factors of $m^2n - 2mn^2$.

The HCF of m^2n and $2mn^2$ is mn.

$\therefore \qquad\qquad m^2n - 2mn^2 = mn(m - 2n)$

$$\text{since}\quad \frac{m^2n}{mn} = m \quad\text{and}\quad \frac{2mn^2}{mn} = 2n$$

c) Find the factors of $3x^4y + 9x^3y^2 - 6x^2y^3$.

The HCF of $3x^4y$, $9x^3y^2$ and $6x^2y^3$ is $3x^2y$.

$\therefore \qquad 3x^4y + 9x^3y^2 - 6x^2y^3 = 3x^2y(x^2 + 3xy - 2y^2)$

$$\text{since}\quad \frac{3x^4y}{3x^2y} = x^2,\quad \frac{9x^3y^2}{3x^2y} = 3xy \quad\text{and}\quad \frac{6x^2y^3}{3x^2y} = 2y^2$$

d) Find the factors of $\dfrac{ac}{x} + \dfrac{bc}{x^2} - \dfrac{cd}{x^3}$

The HCF of $\dfrac{ac}{x}$, $\dfrac{bc}{x^2}$ and $\dfrac{cd}{x^3}$ is $\dfrac{c}{x}$.

$$\therefore \qquad \frac{ac}{x} + \frac{bc}{x^2} - \frac{cd}{x^3} = \frac{c}{x}\left(a + \frac{b}{x} - \frac{d}{x^2}\right)$$

$$\text{since}\quad \frac{ac}{x} \div \frac{c}{x} = a,\quad \frac{bc}{x^2} \div \frac{c}{x} = \frac{b}{x} \quad\text{and}\quad \frac{cd}{x^3} \div \frac{c}{x} = \frac{d}{x^2}$$

Exercise 3.3

Find the HCF of the following:

1) $p^3q^2,\ p^2q^3,\ p^2q$ \qquad\qquad 2) $a^2b^3c^3,\ a^3b^3,\ ab^2c^2$

3) $3mn^2,\ 6mnp,\ 12m^2np^2$ \qquad 4) $2ab,\ 5b,\ 7ab^2$

5) $3x^2yz,\ 12x^2yz,\ 6xy^2z^3,\ 3xyz^2$

Factorise the following:

6) $2x + 6$ \qquad 7) $4x - 4y$ \qquad 8) $5x - 5$

9) $4x - 8xy$ \qquad 10) $mx - my$ \qquad 11) $ax + bx + cx$

12) $\dfrac{x}{2} - \dfrac{y}{8}$ \qquad 13) $5a - 10b + 15c$ \qquad 14) $ax^2 + ax$

15) $2\pi r^2 + \pi rh$ \qquad 16) $3y - 9y^2$ \qquad 17) $ab^3 - a^2b$

18) $x^2y^2 - axy + bxy^2$ \qquad\qquad 19) $5x^3 - 10x^2y + 15xy^2$

20) $9x^3y - 6x^2y^2 + 3xy^5$ \qquad\quad 21) $I_0 + I_0\alpha t$

22) $\dfrac{x}{3} - \dfrac{y}{6} + \dfrac{z}{9}$ \qquad\qquad 23) $2a^2 - 3ab + b^2$

24) $x^3 - x^2 + 7x$ \qquad\qquad 25) $\dfrac{m^2}{pn} - \dfrac{m^3}{pn^2} + \dfrac{m^4}{p^2n^2}$

Factorising by grouping

To factorise the expression $ax + ay + bx + by$ first group the terms in pairs so that each pair of terms has a common factor. Thus,

$$ax + ay + bx + by = (ax + ay) + (bx + by) = a(x + y) + b(x + y)$$

Now notice that in the two terms $a(x + y)$ and $b(x + y)$, $(x + y)$ is a common factor.

Hence, $a(x + y) + b(x + y) = (x + y)(a + b)$

∴ $ax + ay + bx + by = (x + y)(a + b)$

Similarly,

$$np + mp - qn - qm = (np + mp) - (qn + qm)$$
$$= p(n + m) - q(n + m)$$
$$= (n + m)(p - q)$$

Exercise 3.4

Factorise the following:

1) $ax + by + bx + ay$

2) $mp + np - mq - nq$

3) $a^2c^2 + acd + acd + d^2$

4) $2pr - 4ps + qr - 2qs$

5) $4ax + 6ay - 4bx - 6by$

6) $ab(x^2 + y^2) - cd(x^2 + y^2)$

7) $mn(3x - 1) - pq(3x - 1)$

8) $k^2l^2 - mnl - k^2l + mn$

Factorising quadratic expressions

A quadratic expression is one in which the highest power of the symbol used is the square. Typical examples are $x^2 - 5x + 3$ or $3x^2 - 9$ in which there is no power of x greater than x^2.

You will see, from the work in the previous section, that when two binomial expressions are multiplied together the result is always a quadratic expression.

It is often necessary to try and reverse this procedure. This means that we start with a quadratic expression and wish to express this as the product of two binomial expressions—this is not always possible. For example the expressions $x^2 + 1$ or $a^2 + b^2$ cannot be factorised. You may check this for yourself after following the next section of this chapter.

Consider $(7x + 4)(2x + 3) = 14x^2 + 21x + 8x + 12$
$$= 14x^2 + 29x + 12$$

The following points should be noted:

1) The first terms in each bracket when multiplied together give the first term of the quadratic expression.

2) The middle term of the quadratic expression is formed by multiplying together the terms connected by a line (see above equation) and then adding them together.

3) The last terms in each bracket when multiplied together give the last term of the quadratic expression.

In most cases, when factorising a quadratic expression, we find all the possible factors of the first and last terms. Then, by trying various combinations, the combination which gives the correct middle term may be found.

EXAMPLE 3.8
Factorise $2x^2 + 5x - 3$

Factors of $2x^2$		Factors of -3	
$2x$	x	-3	$+1$
		$+3$	-1

Combinations of these factors are:

$(2x - 3)(x + 1) = 2x^2 - x - 3$ which is incorrect,
$(2x + 1)(x - 3) = 2x^2 - 5x - 3$ which is incorrect,
$(2x + 3)(x - 1) = 2x^2 + x - 3$ which is incorrect,
$(2x - 1)(x + 3) = 2x^2 + 5x - 3$ which is correct.

Hence $2x^2 + 5x - 3 = (2x - 1)(x + 3)$

EXAMPLE 3.9
Factorise $12x^2 - 35x + 8$

Factors of $12x^2$		Factors of 8	
$12x$	x	1	8
$6x$	$2x$	8	1
$3x$	$4x$	-1	-8
		-8	-1
		2	4
		4	2
		-2	-4
		-4	-2

By trying each combination in turn the only one which will produce the correct middle term of $-35x$ is found to be $(3x - 8)(4x - 1)$.

$\therefore$ $12x^2 - 35x + 8 = (3x - 8)(4x - 1)$

Where the factors form a perfect square

A quadratic expression, which factorises into the product of two identical brackets resulting in a perfect square, may be factorised by the method used previously. However, if you can recognise that the result will be a perfect square then the problem becomes easier.

It has been shown that

$$(a+b)^2 = a^2 + 2ab + b^2 \quad \text{and} \quad (a-b)^2 = a^2 - 2ab + b^2$$

The square of a binomial expression therefore consists of:

(Square of 1st term) + (Twice product of terms) + (Square of 2nd term)

EXAMPLE 3.10

Factorise $9a^2 + 12ab + 4b^2$

Now $\quad 9a^2 = (3a)^2, \quad 4b^2 = (2b)^2 \quad \text{and} \quad 12ab = 2 \times 3a \times 2b$

$\therefore \qquad\qquad\qquad 9a^2 + 12ab + 4b^2 = (3a+2b)^2$

EXAMPLE 3.11

Factorise $16m^2 - 40m + 25$

Now $16m^2 = (4m)^2, \quad 25 = (-5)^2 \quad \text{and} \quad -40m = 2 \times 4m \times (-5)$

$\therefore \qquad\qquad\qquad 16m^2 - 40m + 25 = (4m-5)^2$

The factors of the difference of two squares

It has previously been shown that
$$(a+b)(a-b) = a^2 - b^2$$

The factors of the difference of two squares are therefore the sum and the difference of the square roots of each of the given terms.

EXAMPLE 3.12

Factorise $9m^2 - 4n^2$

Now $\qquad\qquad\qquad 9m^2 = (3m)^2, \quad \text{and} \quad 4n^2 = (2n)^2$

$\therefore \qquad\qquad\qquad 9m^2 - 4n^2 = (3m+2n)(3m-2n)$

EXAMPLE 3.13

Factorise $4x^2 - 9$

Now $\qquad\qquad\qquad 4x^2 = (2x)^2, \quad \text{and} \quad 9 = (3)^2$

$\therefore \qquad\qquad\qquad 4x^2 - 9 = (2x+3)(2x-3)$

Exercise 3.5

Factorise:

1) $x^2 + 4x + 3$ 2) $x^2 + 6x + 8$

3) $x^2 - 3x + 2$ 4) $x^2 + 2x - 15$

5) $x^2 + 6x - 7$ 6) $x^2 - 5x - 14$

7) $x^2 - 2xy - 3y^2$ 8) $2x^2 + 13x + 15$

9) $3p^2 + p - 2$ 10) $4x^2 - 10x - 6$

11) $3m^2 - 8m - 28$ 12) $21x^2 + 37x + 10$

13) $10a^2 + 19a - 15$ 14) $6x^2 + x - 35$

15) $6p^2 + 7pq - 3q^2$ 16) $12x^2 - 5xy - 2y^2$

17) $x^2 + 2xy + y^2$ 18) $4x^2 + 12x + 9$

19) $p^2 + 4pq + 4q^2$ 20) $9x^2 + 6x + 1$

21) $m^2 - 2mn + n^2$ 22) $25x^2 - 20x + 4$

23) $x^2 - 4x + 4$ 24) $m^2 - n^2$

25) $4x^2 - y^2$ 26) $9p^2 - 4q^2$

27) $x^2 - 1/9$ 28) $1 - b^2$

29) $1/x^2 - 1/y^2$ 30) $121p^2 - 64q^2$

Lowest common multiple (LCM)

The LCM of a set of algebraic terms is the simplest expression of which each of the given terms is a factor.

The method used is similar to that for finding the LCM of a set of numbers.

EXAMPLE 3.14
Find the LCM of $2a$, $3ab$ and a^2b.

We express each term as a product of its factors.

Thus $2a = 2 \times a$

and $3ab = 3 \times a \times b$

and $a^2b = a \times a \times b$

We now note the greatest number of times each factor occurs in any one particular line.

Now   factor 2 occurs once in the line for $2a$,

and   factor 3 occurs once in the line for $3ab$,

and   factor a occurs twice in the line for a^2b,

and   factor b occurs once in either of the lines for $3ab$ or a^2b.

The product of these factors gives the required LCM.

Thus $\qquad$ LCM $= 2 \times 3 \times a \times a \times b$

$$= 6a^2b$$

EXAMPLE 3.15

Find the LCM of $4x$, $8yz$, $2x^2y$ and yz^2.

With practice the LCM may be found by inspection, by finding the product of the highest powers of *all* factors which occur in *any* of the terms.

Thus $\qquad$ LCM $= 8 \times x^2 \times y \times z^2$

$$= 8x^2yz^2$$

EXAMPLE 3.16

Find the LCM of $(a-1)$, $n(m+n)$, $(m+n)^2$.

Brackets must be treated as single factors—*not* the individual terms inside each bracket.

Hence $\qquad$ LCM $= (a-1) \times n \times (m+n)^2$

$$= n(a-1)(m+n)^2$$

Exercise 3.6

Find the LCM for the terms in each of the following examples:

1) $2a$, $3a^2$, a, a^2
2) xy, x^2y, $2x$, $2y$
3) m^2n, mn^2, mn, m^2n^2
4) $2ab$, abc, bc^2
5) $2(x+1)$, $(x+1)$
6) $(a+b)$, $x(a+b)^2$, x^2
7) $(a+b)$, $(a-b)$
8) x, $(1-x)$, $(x+1)$

Algebraic fractions

Since algebraic expressions contain symbols (or letters) which represent numbers, all the rules of operations with numbers also apply to algebraic terms, including fractions.

Thus

$$\frac{1}{\dfrac{1}{a}} = 1 \div \frac{1}{a} = 1 \times \frac{a}{1} = \frac{1 \times a}{1} = a$$

and

$$\frac{\dfrac{a}{b}}{\dfrac{c}{d}} = \frac{a}{b} \div \frac{c}{d} = \frac{a}{b} \times \frac{d}{c} = \frac{a \times d}{b \times c} = \frac{ad}{bc}$$

$$\frac{x+y}{\dfrac{1}{x-y}} = \frac{(x+y)}{\dfrac{1}{(x-y)}} = (x+y) \div \frac{1}{(x-y)} = (x+y) \times \frac{(x-y)}{1}$$

$$= (x+y)(x-y)$$

You should note in the last example how we put bracket round the $x+y$ and $x-y$ to remind us that they must be treated as single expressions—otherwise we may have been tempted to handle the terms x and y on their own.

Adding and subtracting algebraic fractions

Consider the expression $\dfrac{a}{b} + \dfrac{c}{d}$ which is the addition of two fractional terms. These are called partial fractions.

If we wish to express the sum of these partial fractions as one single fraction then we proceed as follows. (The method is similar to that used when adding or subtracting number fractions.)

First find the lowest common demoninator. This is the LCM of b and d which is bd. Each fraction is then expressed with bd as the denominator.

Now $\qquad \dfrac{a}{b} = \dfrac{a \times d}{b \times d} = \dfrac{ad}{bd} \qquad$ and $\qquad \dfrac{c}{d} = \dfrac{c \times b}{d \times b} = \dfrac{cb}{bd}$

and adding these new fractions we have:

$$\frac{a}{b} + \frac{c}{d} = \frac{ad}{bd} + \frac{cb}{bd} = \frac{ad+cb}{bd}$$

EXAMPLE 3.17
Express each of the following as a single fraction:

a) $\dfrac{1}{x} - \dfrac{1}{y}$
b) $a - \dfrac{1}{b}$
c) $\dfrac{1}{m} + n - \dfrac{a}{b}$

d) $\dfrac{a}{b^2} - \dfrac{1}{bc}$
e) $\dfrac{2}{x} + \dfrac{3}{x-1}$

a) $\qquad \dfrac{1}{x} - \dfrac{1}{y} = \dfrac{y}{xy} - \dfrac{x}{xy} \qquad$ since LCM of denominator is xy

$$= \frac{y-x}{xy}$$

b) $\quad a - \dfrac{1}{b} = \dfrac{a}{1} - \dfrac{1}{b}$

$\qquad\qquad = \dfrac{ab}{b} - \dfrac{1}{b}$ $\qquad$ since LCM of denominator is b

$\qquad\qquad = \dfrac{ab - 1}{b}$

c) $\quad \dfrac{1}{m} + n - \dfrac{a}{b} = \dfrac{1}{m} + \dfrac{n}{1} - \dfrac{a}{b}$

$\qquad\qquad = \dfrac{b}{mb} + \dfrac{nmb}{mb} - \dfrac{am}{mb}$ $\qquad$ since LCM m of
$\qquad\qquad\qquad\qquad\qquad\qquad\qquad\qquad$ denominator is mb

$\qquad\qquad = \dfrac{b + nmb - am}{mb}$

d) $\quad \dfrac{a}{b^2} - \dfrac{1}{bc} = \dfrac{ac}{b^2c} - \dfrac{b}{b^2c}$ $\qquad$ since LCM of denominator is b^2c

$\qquad\qquad = \dfrac{ac - b}{b^2c}$

e) $\quad \dfrac{2}{x} + \dfrac{3}{(x-1)} = \dfrac{2(x-1)}{x(x-1)} + \dfrac{3x}{x(x-1)}$ $\qquad$ since LCM of
$\qquad\qquad\qquad\qquad\qquad\qquad\qquad\qquad\qquad$ denominator is $x(x-1)$

$\qquad\qquad = \dfrac{2(x-1) + 3x}{x(x-1)}$

$\qquad\qquad = \dfrac{2x - 2 + 3x}{x(x-1)}$

$\qquad\qquad = \dfrac{5x - 2}{x(x-1)}$

Exercise 3.7

Rearrange the following and thus express in a simplified form:

1) $\quad \dfrac{1}{\frac{b}{a}}$ $\qquad$ 2) $\quad \dfrac{\frac{1}{a}}{\frac{1}{b}}$ $\qquad$ 3) $\quad \dfrac{\frac{x}{y}}{\frac{y}{x}}$ $\qquad$ 4) $\quad \dfrac{1}{\frac{2}{xy}}$

5) $\quad \dfrac{\frac{a}{b}}{a^2}$ $\qquad$ 6) $\quad \dfrac{(a+b)}{\frac{1}{c}}$ $\qquad$ 7) $\quad \dfrac{1-x}{\frac{1}{1+x}}$ $\qquad$ 8) $\quad \dfrac{\frac{1}{a-b}}{\frac{a-b}{c}}$

Express with a common denominator:

9) $\dfrac{1}{x} + \dfrac{1}{y}$

10) $1 + \dfrac{1}{a}$

11) $\dfrac{m}{n} - 1$

12) $\dfrac{b}{c} - c$

13) $\dfrac{a}{b} - \dfrac{c}{d}$

14) $\dfrac{a}{b} - \dfrac{1}{bc}$

15) $\dfrac{1}{zy} + \dfrac{1}{x} + 1$

16) $\dfrac{3}{x} + \dfrac{x}{4}$

17) $\dfrac{3}{c} + \dfrac{2}{d} - \dfrac{5}{e}$

18) $\dfrac{a}{b} + \dfrac{c}{d} + 1$

19) $\dfrac{1}{3fg} - \dfrac{5}{6gh} - \dfrac{1}{2fh}$

20) $\dfrac{2}{x} - \dfrac{4}{x+2}$

21) $1 - \dfrac{x}{x-2}$

Expressing a single fraction as partial fractions

Here we are starting with a single fraction and then splitting it up into partial fractions—in fact it is the exact reverse of adding and subtracting fractions.

The procedure is to take each individual term of the numerator in turn, divide it by the denominator, and then write it on its own.

The following examples use this method.

EXAMPLE 3.18

Express as partial fractions: a) $\dfrac{ab + bc - 1}{abc}$ b) $\dfrac{(x-1) + y}{a(x-1)}$

a)
$$\dfrac{ab + bc - 1}{abc} = \dfrac{ab}{abc} + \dfrac{bc}{abc} - \dfrac{1}{abc}$$
$$= \dfrac{1}{c} + \dfrac{1}{a} - \dfrac{1}{abc}$$

b)
$$\dfrac{(x-1) + y}{a(x-1)} = \dfrac{(x-1)}{a(x-1)} + \dfrac{y}{a(x-1)}$$
$$= \dfrac{1}{a} + \dfrac{y}{a(x-1)}$$

Exercise 3.8

Express as partial fractions:

1) $\dfrac{a+b}{a}$

2) $\dfrac{a-b}{ab}$

3) $\dfrac{1+c}{c}$

4) $\dfrac{x^2 + y}{2x}$

5) $\dfrac{a^2 - ab + ac}{abc}$

6) $\dfrac{x + (x-y)}{x(x-y)}$

Mixed operations with algebraic fractions

We will now combine all the ideas already used. It helps to work methodically and to avoid taking short cuts by leaving out stages of simplification.

EXAMPLE 3.19

Simplify :
    a) $\dfrac{\dfrac{1}{x}+x}{\dfrac{1}{x}}$
    b) $\dfrac{1}{\dfrac{1}{c}-\dfrac{1}{d}}$

a) $\dfrac{\dfrac{1}{x}+x}{\dfrac{1}{x}} = \left(\dfrac{1}{x}+x\right) \div \dfrac{1}{x}$

$= \left(\dfrac{1}{x}+\dfrac{x^2}{x}\right) \times \dfrac{x}{1}$ we are making each term inside the brackets have a common denominator, namely, x

$= \left(\dfrac{1+x^2}{\cancel{x}}\right) \times \cancel{x}$ making use of the common denominator

$= (1+x^2)$ or just $1+x^2$

b) $\dfrac{1}{\dfrac{1}{c}-\dfrac{1}{d}} = 1 \div \left(\dfrac{1}{c}-\dfrac{1}{d}\right)$

$= 1 \div \left(\dfrac{d}{cd}-\dfrac{c}{cd}\right)$ again we are making each term inside the bracket have a common denominator — here cd

$= 1 \div \left(\dfrac{d-c}{cd}\right)$ making use of the common denominator

$= 1 \times \left(\dfrac{cd}{d-c}\right)$

$= \dfrac{cd}{d-c}$ and we cannot simplify any more

Exercise 3.9

Simplify:

1) $\dfrac{1}{1+\dfrac{1}{x}}$

2) $\dfrac{x-\dfrac{1}{x}}{\dfrac{1}{x}}$

3) $\dfrac{\dfrac{1}{a}}{a-\dfrac{1}{a}}$

4) $\dfrac{1}{\dfrac{1}{u}-1}$

5) $\dfrac{\dfrac{1}{x}+\dfrac{1}{y}}{\dfrac{1}{xy}}$

6) $\dfrac{1}{\dfrac{1}{R_1}+\dfrac{1}{R_2}}$

4 Solution of equations

Linear equations occur more frequently than most other types: they are equations in which the variable is no higher than the first power, e.g. x or p. Quadratic equations contain a variable to the second power, e.g. x^2 or p^2, but no higher. These types of equations arise from the modelling of engineering situations and so we must be able to solve them and find numerical solutions. Simultaneous equations result from a problem containing two sets of data which must be taken together when solving. For instance the use of Kirchoff's at junctions in an electric circuit network may result in two relationships which are both two completely different equations, each containing the same two unknown currents. You may wonder why transposition of equations (which forms part of PC3) has been included here and not in Chapter 3. Well the techniques used in solving linear equations are also those used for transposing equations, and we think that you will find them easier to follow in this sequence.

Linear equations

Linear equations contain only the first power of the unknown quantity. Remember the first power of anything is written as itself, e.g. $a^1 = a$

So $\qquad x + 7 = 10, \quad 6t = 4t + 9 \quad$ and $\quad \dfrac{5p}{3} = \dfrac{2p + 6}{2}$

are all examples of linear equations.

Solving equations

The statement ‘a number plus 7 equals 10’

may be written as an equation $x + 7 = 10$

We can see that there is only one value of x, namely 3, which will ‘satisfy’ the equation, or make the left-hand side (LHS) equal to the right-hand side (RHS). The process of finding $x = 3$ is called ‘solving’ the equation, and the value 3 is known as the ‘solution’ or ‘root’ of the equation.

When handling equations balance must be maintained between the LHS and the RHS.

Consider the pair of scales in balance.

A mathematical model of this arrangement is the equation

$x = 3$ [1]

> Now whatever changes we make to the left hand and right hand sides of an equation, *balance* must be maintained.

This new loading still keeps the balance.

Now multiplying both sides of the equation [1] by 4, then

$4x = 4 \times 3$

giving

$4x = 12$

> Remember that when ‘manipulating’ an equation whatever you do to one side of an equation, you must do the same to the other.

For example if we add 7 to both sides of equation [1]

$x + 7 = 3 + 7$

giving

$x + 7 = 10$

Or if we divide both sides of equation [1] by 5 then

$\dfrac{x}{5} = \dfrac{3}{5}$

We may even become ambitious and square both sides of equation [1]

$x^2 = 3^2$

giving

$x^2 = 9$

Now let us use this idea of ‘doing the same thing to both sides’ to solve some equations.

EXAMPLE 4.1

Solve $\qquad\qquad\qquad 7x = 5x + 18$

We shall group all the terms containing the unknown, here x, on one side (the LHS of the given equation), and the remainder on the other.

So if we subtract $5x$ $\quad\rbrace$
from both sides, then $\qquad\qquad 7x - 5x = 5x + 18 - 5x$

giving $\qquad\qquad\qquad\qquad\qquad 2x = 18$

and dividing both sides $\quad\rbrace$
by 2 $\qquad\qquad\qquad\qquad\qquad \dfrac{2x}{2} = \dfrac{18}{2}$

giving $\qquad\qquad\qquad\qquad\qquad x = 9$

We must now check to see if this value is correct.
We shall substitute the value into both sides of the original equation.

Check: When $x = 9$ then LHS $= 7 \times 9 = 63$, RHS $= 5 \times 9 + 18 = 63$. Thus the value obtained is correct.

We say that $x = 9$ is the solution, or root, of the given equation —or we may say that $x = 9$ satisfies the given equation.

EXAMPLE 4.2

Solve $\qquad\qquad\qquad 2(4y + 3) = 3y + 8$

Removing the bracket gives $\qquad\quad 8y + 6 = 3y + 8$

Subtracting $3y$ and 6 $\quad\rbrace$
from both sides, then $\qquad 8y + 6 - 3y - 6 = 3y + 8 - 3y - 6$

giving $\qquad\qquad\qquad\qquad\qquad 5y = 2$

Dividing both sides $\quad\rbrace$
by 5, then $\qquad\qquad\qquad\qquad \dfrac{5y}{5} = \dfrac{2}{5}$

from which $\qquad\qquad\qquad\qquad y = \dfrac{2}{5}$ or $\;0.4$

Check: When $x = 0.4$

LHS $= 2(4 \times 0.4 + 3) = 9.2$ $\rbrace$
RHS $= 3 \times 0.4 + 8 = 9.2$ $\quad$ Hence the solution is correct.

Equations containing fractions

Consider $\qquad \dfrac{2t}{5} + \dfrac{3}{2} = \dfrac{3t}{4} + 6 \qquad$ and $\qquad \dfrac{1}{x-2} = \dfrac{2}{x} - 7$

Both are equations which contain fractions, and they are *not* the 'flat line type' of equation with which we are familiar.

The method we use on these is to eliminate the fractions as a first step. This is achieved by multiplying through (i.e. multiplying both sides of the equation) by the LCM of the denominators.

You should remember that any individual term is a fraction if it has a top and bottom line; so both $\dfrac{3}{2}$ and $\dfrac{1}{x-2}$ are fractions and we must give them similar treatment.

EXAMPLE 4.3

Solve the equation

$$\frac{2t}{5} + \frac{3}{2} = \frac{3t}{4} + 6$$

The LCM of the denominators 5, 2 and 4 is 20, since 20 is the lowest whole number into which 5, 2 and 4 will divide easily.
So multiplying both sides by 20 we have

$$\left(\frac{2t}{5} + \frac{3}{2}\right) \times 20 = \left(\frac{3t}{4} + 6\right) \times 20$$

and removing brackets

$$\frac{2t \times 20}{5} + \frac{3 \times 20}{2} = \frac{3t \times 20}{4} + 6 \times 20$$

giving

$$8t + 30 = 15t + 120$$

Subtracting 15t and 30 from both sides, then $\Big\}$ $\quad 8t + \cancel{30} - 15t - \cancel{30} = \cancel{15t} + 120 - \cancel{15t} - 30$

giving

$$-7t = 90$$

Dividing both sides by − 7, then $\Big\}$

$$\frac{\cancel{-7}t}{\cancel{-7}} = \frac{90}{7}$$

giving

$$t = -\frac{90}{7}$$

or

$$t = -12.9 \quad \text{correct to 3 s.f.}$$

Check: If $t = -12.9$ then $\quad$ LHS $= \frac{2}{5}(-12.9) + \frac{3}{2} = -3.66$
$\quad\quad\quad\quad\quad\quad$ and $\quad$ RHS $= \frac{3}{4}(-12.9) + 6 = -3.68$

Bearing in mind the answer was rounded to 3 s.f. we may say that the LHS = RHS, and hence that the solution is correct.

EXAMPLE 4.4

Solve

$$\frac{z-4}{3} - \frac{2z-1}{2} = 4$$

The LCM of the denominators 3 and 2 is 6.

Multiplying through by 6 $\Big\}$

$$\frac{(z-4)}{3} \times 6 - \frac{(2z-1)}{2} \times 6 = 4 \times 6$$

giving

$$2(z-4) - 3(2z-1) = 24$$

Removing the brackets

$$2z - 8 - 6z + 3 = 24$$

giving

$$-4z - 5 = 24$$

Adding 5 to both sides

$$-4z - \cancel{5} + \cancel{5} = 24 + 5$$

giving

$$-4z = 29$$

Dividing both sides by − 4, then $\Big\}$

$$\frac{\cancel{-4}z}{\cancel{-4}} = \frac{29}{-4}$$

giving

$$z = -\frac{29}{4} \quad \text{or} \quad -7.25$$

Check this value for yourself to see if it is correct.

EXAMPLE 4.5

Solve the equation $\dfrac{3}{m+4} = \dfrac{7}{m}$

In order to bring the unknowns m to 'the top line' we will multiply both sides of the equation by the LCM of the denominators $m(m+4)$.

Multiplying both sides by $m(m+4)$, then $\Big\}$ $\dfrac{3}{(m+4)} \times m(m+4) = \dfrac{7}{m} \times m(m+4)$

giving $\qquad\qquad 3m = 7(m+4)$

$\therefore \qquad\qquad 3m = 7m + 28$

Subtracting $7m$ from both sides, then $\Big\}$ $\qquad 3m - 7m = 7m + 28 - 7m$

giving $\qquad\qquad -4m = 28$

Dividing both sides by -4 $\qquad \dfrac{-4m}{-4} = \dfrac{28}{-4}$

from which $\qquad\qquad m = -7$

Again we will leave you to check if this answer is correct.

Exercise 4.1

1) $6m + 11 = 24 - m$

2) $1.2x - 0.8 = 0.8x + 1.2$

3) $5(m - 2) = 15$

4) $3(x - 1) - 4(2x + 3) = 14$

5) $4(x - 5) = 7 - 5(3 - 2x)$

6) $3x = 5(9 - x)$

7) $\dfrac{x}{5} - \dfrac{x}{3} = 2$

8) $3m + \dfrac{3}{4} = 2 + \dfrac{2m}{3}$

9) $\dfrac{4}{t} = \dfrac{2}{3}$

10) $\dfrac{4}{7}y - \dfrac{3}{5}y = 2$

11) $\dfrac{1}{3x} + \dfrac{1}{4x} = \dfrac{7}{20}$

12) $\dfrac{x+3}{4} - \dfrac{x-3}{5} = 2$

13) $\dfrac{3-u}{4} = \dfrac{u}{3}$

14) $\dfrac{x-2}{x-3} = 3$

15) $\dfrac{3}{v-2} = \dfrac{4}{v+4}$

16) $\dfrac{x}{3} - \dfrac{3x-7}{5} = \dfrac{x-2}{6}$

17) Kirchoff's laws in an electric circuit show that currents i_1, i_2 and i_3 are connected by the equation $2i_1 + 3i_2 + i_3 = 5(i_2 - i_1)$. If $i_2 = 7\,\text{A}$ and $i_3 = 5\,\text{A}$, find the value of current i_1.

18) If the sides of a sheet metal template which is triangular in shape have lengths of l, $(l + 2)$ and $(2l + 3)$, and the perimeter is 200 mm, find the length of the shortest side.

19) An inspection hatch covers an aperture which is rectangular in shape. If the depth, d, of the aperture is one half of the width, and the perimeter is 1.8 metres find the depth in millimetres.

20) Current division in an electric circuit is given by

$$i_1 = \left(\frac{R_2}{R_1 + R_2}\right) i.$$ Find resistance R_2 if $R_1 = 3$ ohms, $i_1 = 6$ amperes and $i = 9$ amperes.

21) In a particular fine measurement we have two lengths x and $(x + 2)$. If the ratio of these lengths is 2:3, what is the value of x?

22) For resistance in parallel the effective resistance R is given by the expression $R = \dfrac{R_1 R_2}{R_1 + R_2}$

Find the value of R_1 if $R = 1.8$ ohms and $R_2 = 4.9$ ohms.

Quadratic equations

Just as we have been able to solve simple equations, we may well have to cope with more difficult arrangements, one such example is a quadratic equation.

An equation of the type $ax^2 + bx + c = 0$, involving x in the second degree and containing no higher power of x, is called a *quadratic equation*. The constants a, b and c have any numerical values. Thus,

$$x^2 - 9 = 0 \quad \text{where } a = 1, b = 0 \text{ and } c = -9$$
$$x^2 - 2x - 8 = 0 \quad \text{where } a = 1, b = -2 \text{ and } c = -8$$
$$2.5x^2 - 3.1x - 2 = 0 \quad \text{where } a = 2.5, b = -3.1 \text{ and } c = -2$$

are all examples of quadratic equations. A quadratic equation may contain only the square of the unknown quantity, as in the first of the above equations, or it may contain both the square and the first power as in the other two.

There are two commonly used methods of solution:

1) by factorisation,
2) by use of a formula.

1) *Factorisation* is the reverse of multiplying two brackets together. On seeing a quadratic equation our first thoughts are usually 'can we factorise it easily'? If we can, it provides a quick and neat method, but if there is any difficulty go immediately to method 2):

2) *Use of a formula* which, although more tedious, can be used for solving any quadratic equation.

Roots of an equation

Before we commence solving equations it is instructive to work backwards from knowing the solutions or roots, and then constructing the quadratic equation which they satisfy.

If either of two factors has zero value, then their product is zero. Thus if either $M = 0$ or $N = 0$ then $M \times N = 0$

Now suppose that either $\qquad x = 1$ or $\qquad x = 2$

$\therefore$ rearranging gives either $\qquad x - 1 = 0$ or $x - 2 = 0$

Hence $\qquad (x - 1)(x - 2) = 0$

since either of the factors has zero value.

If we now multiply out the brackets of this equation we have

$$x^2 - 3x + 2 = 0$$

and we know that $x = 1$ and $x = 2$ are values of x which satisfy this equation. The values 1 and 2 are called the solutions or *roots* of the equation $x^2 - 3x + 2 = 0$

EXAMPLE 4.6
Find the equation whose roots are -2 and 4

From the values given either $\quad x = -2$ or $\qquad x = 4$

$\therefore$ either $\qquad\qquad x + 2 = 0$ or $x - 4 = 0$

Hence $\qquad\qquad (x + 2)(x - 4) = 0$

since either of the factors has zero value.

$\therefore$ Multiplying out gives $\quad x^2 - 2x - 8 = 0$

EXAMPLE 4.7
Find the equation whose roots are 3 and -3

From the values given either $\quad x = 3$ or $\qquad x = -3$

$\therefore$ either $\qquad\qquad x - 3 = 0$ or $\quad x + 3 = 0$

Hence $\qquad\qquad (x - 3)(x + 3) = 0$

since either of the factors has zero value.

Multiplying out we have $\qquad x^2 - 9 = 0$

EXAMPLE 4.8
Find the equation whose roots are 5 and 0.

From the given values given either $\quad x = 5$ or $x = 0$

$\therefore$ either $\qquad\qquad x - 5 = 0$ or $\quad x = 0$

Hence $\qquad\qquad x(x - 5) = 0$

since either of the factors has zero value.

And multiplying out we have $\quad x^2 - 5x = 0$

Exercise 4.2

Find the equations whose roots are:

1)	3, 1	2)	2, −4
3)	−1, −2	4)	1.6, 0.7
5)	2.73, −1.66	6)	−4.76, −2.56
7)	0, 1.4	8)	−4.36, 0
9)	−3.5, +3.5	10)	repeated, each = 4

Solution by Factors

This method is the reverse of the procedure used to find an equation when given the two roots. We shall now start with the equation and proceed to solve the equation and find the roots.

We shall again use the fact that if the product of two factors is zero then one factor or the other must be zero. Thus if $M \times N = 0$ then either $M = 0$ or $N = 0$

When the factors are easy to find the factor method is very quick and simple. However do not spend too long trying to find factors: if they are not easily found use the formula given in the next method (p. 98) to solve the equation.

EXAMPLE 4.9
Solve the equation $(2x + 3)(x - 5) = 0$

Since the product of the two factors $2x + 3$ and $x - 5$ is zero then

either $\qquad\qquad 2x + 3 = 0 \qquad$ or $\quad x - 5 = 0$

Hence $\qquad\qquad x = -\dfrac{3}{2} \quad$ or $\quad x = 5$

EXAMPLE 4.10
Solve the equation $6x^2 + x - 15 = 0$

Factorising gives $(2x - 3)(3x + 5) = 0$

$\therefore$ either $\qquad\qquad 2x - 3 = 0 \quad$ or $\quad 3x + 5 = 0$

Hence $\qquad\qquad x = \dfrac{3}{2} \quad$ or $\quad x = -\dfrac{5}{3}$

EXAMPLE 4.11

Solve the equation $14x^2 = 29x - 12$

Bring all the terms to the left-hand side:

$$14x^2 - 29x + 12 = 0$$

Factorising gives $\qquad (7x - 4)(2x - 3) = 0$

∴ either $\qquad 7x - 4 = 0 \quad$ or $\quad 2x - 3 = 0$

Hence $\qquad x = \dfrac{4}{7} \quad$ or $\quad x = \dfrac{3}{2}$

EXAMPLE 4.12

Find the roots of the equation $x^2 - 16 = 0$

Factorising gives $\qquad (x - 4)(x + 4) = 0$

∴ either $\qquad x - 4 = 0 \quad$ or $\quad x + 4 = 0$

Hence $\qquad x = 4 \quad$ or $\quad x = -4$

In this case an alternative method may be used:

Rearranging the given equation gives $\ x^2 = 16$

and taking the square root of both sides $\ x = \sqrt{16} = \pm 4$

Remember that when we take a square root we must insert the $\pm$ sign, because $\ (+4)^2 = 16 \ $ and $\ (-4)^2 = 16$

EXAMPLE 4.13

Solve the equation $x^2 - 2x = 0$

Factorising gives $\qquad x(x - 2) = 0$

∴ either $\qquad x = 0 \quad$ or $\quad x - 2 = 0$

Hence $\qquad x = 0 \quad$ or $\quad x = 2$

Note: The solution $x = 0$ must not be omitted as it is a solution in the same way as $x = 2$ is a solution. Equations should not be divided through by variables, such as x, since this removes a root of the equation.

EXAMPLE 4.14

Solve the equation $x^2 - 6x + 9 = 0$

Factorising gives $\qquad (x - 3)(x - 3) = 0$

∴ either $\qquad x - 3 = 0 \quad$ or $\quad x - 3 = 0$

Hence $\qquad x = 3 \quad$ or $\quad x = 3$

In this case there is only one arithmetical value for the solution. Technically, however, there are two roots and when they have the same numerical value they are said to be repeated roots.

Solution by Formula

In general, quadratic expressions do not factorise easily and so some other method of solving quadratic equations must be used.

The *standard form* of the *quadratic equation* is:

$$ax^2 + bx + c = 0$$

The *solution* of the equation is:

$$x = \frac{-b \pm \sqrt{b^2 - 4ac}}{2a}$$

EXAMPLE 4.15
Solve the equation $3x^2 - 8x + 2 = 0$

Comparing with $ax^2 + bx + c = 0$, we have $a = 3$, $b = -8$ and $c = 2$

If your calculator has a built-in program for solving quadratic equations, go ahead and make use of it. If not, choose your own calculator sequence for the solution shown below. We would suggest that when you first find the value of the square root you put it into memory so that it can be recalled later when you are finding the other solution.

Substituting these values in the formula, we have

$$x = \frac{-(-8) \pm \sqrt{(-8)^2 - 4 \times 3 \times 2}}{2 \times 3}$$

$$= \frac{8 \pm \sqrt{64 - 24}}{6} = \frac{8 \pm \sqrt{40}}{6} = \frac{8 \pm 6.325}{6}$$

$\therefore$ either $\quad x = \dfrac{8 + 6.325}{6} \quad$ or $\quad x = \dfrac{8 - 6.325}{6}$

The two solutions are therefore

$$x = 2.39 \quad \text{or} \quad x = 0.28 \quad \text{correct to 2 d.p.}$$

It is important that we check the solutions in case we have made an error. We may do this by substituting the values obtained in the left-hand side of the given equation and checking that the solution is zero, or approximately zero.

Thus when $x = 2.39$ we have LHS $= 3(2.39)^2 - 8(2.39) + 2 \approx 0$
and when $x = 0.28$ we have LHS $= 3(0.28)^2 - 8(0.28) + 2 \approx 0$

EXAMPLE 4.16

Solve the equation $\quad 2.13x^2 + 0.75 - 6.89 = 0$

Here $\quad a = 2.13, \, b = 0.75, \, c = -6.89$

$$x = \frac{-0.75 \pm \sqrt{(0.75)^2 - 4(2.13)(-6.89)}}{2 \times 2.13}$$

$$= \frac{-0.75 \pm 7.698}{4.26}$$

$\therefore$ either $\quad x = \dfrac{-0.75 + 7.698}{4.26} \quad$ or $\quad x = \dfrac{-0.75 - 7.698}{4.26}$

Hence $\quad x = 1.631 \quad$ or $\quad x = -1.983 \quad$ correct to 4 s.f.

Solution check

When $\qquad\qquad x = 1.631$

we have $\qquad$ LHS $= 2.13(1.631)^2 + 0.75(1.631) - 6.89 \approx 0$

When $\qquad\qquad x = -1.983$

we have $\qquad$ LHS $= 2.13(-1.983)^2 + 0.75(-1.983) - 6.89 \approx 0$

EXAMPLE 4.17

Solve the equation $\quad x^2 + 4x + 5 = 0$

Here $\quad a = 1, \, b = 4$ and $c = 5$

$$\therefore \quad x = \frac{-4 \pm \sqrt{4^2 - 4(1)(5)}}{2(1)} = \frac{-4 \pm \sqrt{16 - 20}}{2} = \frac{-4 \pm \sqrt{-4}}{2}$$

Now when a number is squared the answer must be a positive quantity because two quantities having the same sign are being multiplied together. Therefore the square root of a negative quantity, as $\sqrt{-4}$ in the above equation, has no arithmetical meaning and is called an imaginary quantity. The equation $x^2 + 4x + 5 = 0$ is said to have imaginary or complex roots. Equations which have complex roots are beyond the scope of this book and are dealt with in more advanced mathematics.

Exercise 4.3

Solve the following equations by the factor method:

1) $x^2 - 36 = 0$

2) $4x^2 - 6.25 = 0$

3) $9x^2 - 16 = 0$

4) $x^2 + 9x + 20 = 0$

5) $x^2 + x - 72 = 0$

6) $3x^2 - 7x + 2 = 0$

7) $m^2 = 6m - 9$

8) $m^2 + 4m + 4 = 36$

9) $14q^2 = 29q - 12$

10) $9x + 28 = 9x^2$

Solve the following equations by using the quadratic formula:

11) $4x^2 - 3x - 2 = 0$

12) $x^2 - x + \frac{1}{4} = \frac{1}{9}$

13) $3x^2 + 7x - 5 = 0$

14) $7x^2 + 8x - 2 = 0$

15) $5x^2 - 4x - 1 = 0$

16) $2x^2 - 7x = 3$

17) $x^2 + 0.3x - 1.2 = 0$

18) $2x^2 - 5.3x + 1.25 = 0$

Solve the following equations:

19) $x(x+4) + 2x(x+3) = 5$

20) $x^2 - 2x(x-3) = -20$

21) $\dfrac{2}{x+2} + \dfrac{3}{x+1} = 5$

22) $\dfrac{x+2}{3} - \dfrac{5}{x+2} = 4$

23) $\dfrac{6}{x} - 2x = 2$

24) $40 = \dfrac{x^2}{80} + 4$

25) $\dfrac{x+2}{x-2} = x - 3$

26) $\dfrac{1}{x+1} - \dfrac{1}{x+3} = 15$

Probelms involving quadratic equations

Having learnt how to solve quadratic equations, we shall now see how they arise from engineering situations.

EXAMPLE 4.18
The distance, $s\,$m, moved by a vehicle in time, $t\,$s, with an initial velocity, $v_1\,$m/s, and a constant acceleration, $a\,$m/s^2, is given by $s = v_1 t + \frac{1}{2}at^2$. Find the time taken to cover 84 m with a constant acceleration 2 m/s^2 if the initial velocity is 5 m/s.

Using $s = 84\,$m, $v_1 = 5\,$m/s and $a = 2\,$m/s^2 we have

$$84 = 5t + \tfrac{1}{2}2t^2$$

from which $\qquad\qquad t^2 + 5t - 84 = 0$

Factorising gives $\qquad (t + 12)(t - 7) = 0$

$\therefore$ either $\qquad\qquad\qquad t + 12 = 0 \qquad$ or $\quad t - 7 = 0$

$\therefore$ either $\qquad\qquad\qquad\qquad t = -12 \quad$ or $\qquad t = 7$

Now the solution $t = -12$ is not acceptable since negative time has no meaning in this question. Thus the required time is 7 seconds.

Solution check
When $t = 7$ we have LHS $= 7^2 + 5 \times 7 - 84 = 0$

EXAMPLE 4.19
A miniaturised circuit board is rectangular in shape.

The diagonal of the rectangle is 15 mm long and one side is 2 mm longer than the other. Find the dimensions of the circuit board.

In Fig. 4.1, let the length of BC be $x\,$mm. The length of CD is then $(x + 2)\,$mm. $\triangle$BCD is right-angled and so by Pythagoras,

$$x^2 + (x+2)^2 = 15^2$$

$\therefore \qquad\qquad\qquad x^2 + x^2 + 4x + 4 = 225$

$\therefore \qquad\qquad\qquad 2x^2 + 4x - 221 = 0$

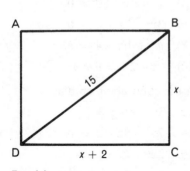

Fig. 4.1

Here $a = 2$, $b = 4$ and $c = -221$

$$\therefore \qquad x = \frac{-4 \pm \sqrt{4^2 - 4 \times 2 \times (-221)}}{2 \times 2}$$

$$\therefore \qquad x = 9.56 \quad \text{or} \quad -11.56 \qquad \text{correct to 2 d.p.}$$

Since the answer cannot be negative, then $x = 9.56 \, \text{mm}$

Now $\qquad\qquad\qquad\qquad x + 2 = 11.56 \, \text{mm}$

$\therefore$ The circuit board has adjacent sides equal to 9.56 mm and 11.56 mm.

Solution check
When $x = 9.56$ we have LHS $= 2(9.56)^2 + 4(9.56) - 221 \approx 0$

EXAMPLE 4.20
A section of an air duct is shown by the full lines in Fig. 4.2.

a) Show that: $\quad w^2 - 2Rw + \dfrac{R^2}{4} = 0$

b) Find the value of w when $R = 2 \, \text{m}$.

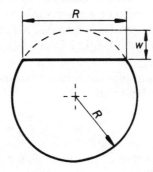

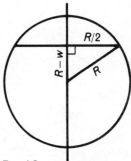

Fig. 4.2 $\qquad\qquad\qquad\qquad$ Fig. 4.3

a) Using the construction shown in Fig. 4.3 we have, by Pythagoras,

then $\qquad\qquad\qquad\qquad (R - w)^2 + \left(\dfrac{R}{2}\right)^2 = R^2$

$$\therefore \qquad\qquad R^2 - 2Rw + w^2 + \frac{R^2}{4} = R^2$$

$$\therefore \qquad\qquad w^2 - 2Rw + \frac{R^2}{4} = 0$$

b) When $R = 2$, $\qquad\qquad w^2 - 4w + 1 = 0$

Here $a = 1$, $b = -4$ and $c = 1$

$$\therefore \qquad w = \frac{-(-4) \pm \sqrt{(-4)^2 - 4 \times 1 \times 1}}{2 \times 1}$$

Hence $\qquad w = 3.732 \quad \text{or} \quad 0.268 \, \text{m} \qquad \text{correct to 3 d.p.}$

Now w must be less than 2 m, thus $w = 0.268 \, \text{m}$

Solution check
When $w = 0.268$ we have LHS $= (0.268)^2 - 4(0.268) + 1 \approx 0$

Exercise 4.4

1) The length L of a wire stretched tightly between two supports in the same horizontal line is given by

$$L = S + \frac{8D^2}{3S}$$

where S is the span and D is the (small) sag. If $L = 150$ and $D = 5$, find the value of S.

2) The profile of a pattern for a sheet metal infill plate is that of a right-angled triangle. The longest side is twice as long as one of the sides forming the right angle. If the remaining side is 80 mm long, find the length of the longest side.

3) A rectangular shutter for closing off a ventilator duct covers an area of 6175 mm². If its width is 30 mm greater than its depth, find its dimensions.

4) A closed cylindrical container is used as part of a test rig, and will eventually become a reservoir in an expansion circuit for a cooling liquid. If its total outer surface area is 29 000 mm², find the value of its radius if the height of the container is 75 mm.

5) If a segment of a circle has a radius R, a height H and a length of chord W show that

$$R = \frac{W^2}{8H} + \frac{H}{2}$$

Rearrange this equation to give a quadratic equation for H and hence find H when $R = 12\,\mathrm{m}$ and $W = 8\,\mathrm{m}$.

6) Fig. 4.4 shows a template whose area is 9690 mm². Find the value of r.

7) A pressure vessel is of the shape shown in Fig. 4.5, the radius of the vessel being r mm. If the surface area is 30 000 mm³ find r.

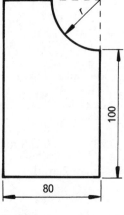

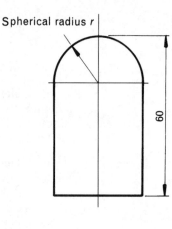

Spherical radius r

Fig. 4.4   Fig. 4.5

8) The total iron loss in a transformer is given by the equation $P = 0.1f + 0.006f^2$. If $P = 20$ watts, find the value of the frequency f.

9) The volume of a frustum of a cone is given by the formula $V = \frac{1}{3}(R^2 + rR + r^2)$ where h is the height of the frustum and R and r are the radii at the large and small ends respectively. If $h = 9$ m, $R = 4$ m and the volume is 337.2 m^3, what is the value of r?

10) A square steel plate is pierced by a square tool leaving a margin of 20 mm all round. The area of the hole is one third that of the original plate. What are the dimensions of the original plate?

11) The velocity, v, of a body in terms of time, t, is given by the expression $v = 3t^2 - 6t - 3$. Find the times at which the velocity is zero.

Simultaneous equations

Solution of simultaneous linear equations

Remember that linear equations are those which contain only the first power of the unknown quantities. Two such equations may be obtained by applying Kirchoff's laws to two junctions in an electric circuit network. These equations both contain unknown currents i_1 and i_2 and have to be solved 'together', or 'simultaneously' as we say.

EXAMPLE 4.21

Two equations for currents i_1 and i_2 obtained as explained above are $5i_1 + 3i_2 = 19$ and $3i_1 + 2i_2 = 12$. Solve these equations simultaneously and hence find i_1 and i_2.

We have	$5i_1 + 3i_2 = 19$	[1]
and	$3i_1 + 2i_2 = 12$	[2]

Now if we multiply equation [1] by 3, and equation [2] by 5, then the coefficient of i_1 will be the same in both equations.

Thus	$15i_1 + 9i_2 = 57$	[3]
and	$15i_1 + 10i_2 = 60$	[4]

We may now eliminate i_1 by subtracting equation [3] from equation [4],

giving	$10i_2 - 9i_2 = 60 - 57$
or	$i_2 = 3$

To find the other unknown i_1 we may substitute in either of the original equations. So substituting $i_2 = 3$ into equation [1],

then	$5i_1 + 3 \times 3 = 19$
giving	$5i_1 = 19 - 9$
and hence	$i_1 = 2$

Thus the two currents are $i_1 = 2$ and $i_2 = 3$.

We must now check these results. We know that the values satisfy equation [1] since it was used in the solution, so we should try the values in equation [2].

Thus for equation [2] $\text{LHS} = 3 \times 2 + 2 \times 3$

$$= 12 = \text{RHS}$$

So the values $i_1 = 2$ and $i_2 = 3$ also satisfy equation [2], and therefore may be considered to be correct.

Exercise 4.5

1) $3x + 2y = 7$
   $x + y = 3$

2) $4x - 3y = 1$
   $x + 3y = 19$

3) $x + 3y = 7$
   $2x - 2y = 6$

4) $7x - 4y = 37$
   $6x + 3y = 51$

5) $4x - 6y = -2.5$
   $7x - 5y = -0.25$

6) $x + y = 17$
   $\dfrac{x}{5} - \dfrac{y}{7} = 1$

7) $\dfrac{3x}{2} - 2y = \dfrac{1}{2}$
   $x + \dfrac{3y}{2} = 6$

8) $2x + \dfrac{y}{2} = 11$
   $\dfrac{3x}{5} + 3y = 9$

Problems involving simultaneous equations

In problems which involve two unknowns it is necessary to form two separate equations from the given data and then to solve these as shown above.

EXAMPLE 4.22

In a certain lifting machine it is found from a test that the effort (E) and the load (W) which is being raised are connected by the equation $E = aW + b$. An effort of 3.7 N raises a load of 10 N whilst an effort of 7.2 N raises a load of 20 N. Find the values of the constants a and b and hence find the effort needed to lift a load of 12 N.

Substituting $E = 3.7$ and $W = 10$ into the given equation we have

$$3.7 = 10a + b \qquad\qquad [1]$$

Substituting $E = 7.2$ and $W = 20$ into the given equation we have

$$7.2 = 20a + b \qquad\qquad [2]$$

Subtracting equation [1] from equation [2] gives

$$3.5 = 10a$$
$$a = 0.35$$

Substituting for a in equation [1] gives

$$3.7 = 10 \times 0.35 + b$$
$$3.7 = 3.5 + b$$
$$3.7 - 3.5 = b$$
$$b = 0.2$$

The given equation therefore becomes:

$$E = 0.35W + 0.2$$

We leave you to check if these answers are correct.

When $\quad W = 12 \qquad E = 0.35 \times 12 + 0.2 = 4.2 + 0.2 = 4.4$ N

Hence an effort of 4.4 N is needed to raise a load of 12 N.

EXAMPLE 4.23

A heating installation for one house consists of 5 radiators and 4 convector heaters and the cost of the installation is £1130. In a second house 6 radiators and 7 convector heaters are used, the cost of this installation being £1740. The labour costs are £400 and £600, respectively. Find the cost of a radiator and the cost of a convector heater.

For the first house the cost of the heaters is

$$£1130 - £400 = £730$$

For the second house the cost of the heaters is

$$£1740 - £600 = £1140$$

let £x be the cost of a radiator and £y be the cost of a convector heater.

For the first house,	$5x + 4y = 730$	[1]
For the second house,	$6x + 7y = 1140$	[2]
Multiplying [1] by 6 gives	$30x + 24y = 4380$	[3]
Multiplying [2] by 5 gives	$30x + 35y = 5700$	[4]

Subtracting equation [3] from equation [4] gives

$$11y = 1320$$
$$y = 120$$

Substituting for y in equation [1] gives

$$5x + 4 \times 120 = 730$$
$$5x = 250$$
$$x = 50$$

Therefore the cost of a radiator is £50 and the cost of a convector heater is £120.

You should check these results as before.

Exercise 4.6

1) In an experiment to find the friction force F between two metallic surfaces when the load is W, the law connecting the two quantities was of the type $F = mW + b$. When $F = 2.5$, $W = 6$ and when $F = 3.1$, $W = 9$. Find the values of m and b. Hence find the value of F when $W = 12$.

2) A foreman and seven men together earn £1400 per week whilst two foremen and 17 men together earn £3304 per week. Find the earnings for a foreman and for a man.

3) For one installation 6 ceiling roses and 8 plugs are required, the total cost of these items being £16. For a second installation 5 ceiling roses and 12 plugs are used, the cost being £17.60. Find the cost of a ceiling rose and a plug.

4) In a certain lifting machine it is found that the effort, E, and the load, W, are connected by the equation $E = aW + b$. An effort of 2.6 raises a load of 8, whilst an effort of 3.8 raises a load of 12. Find the values of the constants a and b and determine the effort required to raise a load of 15.

5) If 100 m of wire and 8 plugs cost £62 and 150 m of wire and 10 plugs cost £90, find the cost of 1 m of wire and the cost of a plug.

6) An alloy containing $8\,\text{cm}^3$ of copper and $7\,\text{cm}^3$ of tin has a mass of 121 g. A second alloy containing $9\,\text{cm}^3$ of copper and 11cm^3 of tin has a mass of 158 g. Find the densities of copper and tin in g/cm^3.

7) An equation of motion for a motor vehicle is $s = ut + \frac{1}{2}at^2$ where $s\,\text{m}$ is the distance travelled at time $t\,\text{s}$, $u\,\text{ms}^{-1}$ is the initial velocity and $a\,\text{ms}^{-2}$ is the constant acceleration.

   Test measurements give $s = 32\,\text{m}$ when $t = 1.5\,\text{s}$
   and $s = 71\,\text{m}$ when $t = 3\,\text{s}$

   Find the values of the initial velocity and of the constant acceleration.

8) Applying Kirchoff's laws to an electric circuit gives the following two relationships for currents i_1 and i_2:

$$6 - 0.3i_1 = 2(i_1 - i_2)$$

   and $$12 - 3.4i_2 = 3(i_2 - i_1)$$

   Find the values of i_1 and i_2.

9) The relationship $R = R_0(1 + \alpha t)$ shows how a resistance R ohms varies with temperature $t°C$. R_0 is the value of the resistance at $0°C$, and α is called the temperature coefficient of resistance. If $R = 28\,\text{ohms}$ at $40°C$, and $R = 34\,\text{ohms}$ at $100°C$, find the values of R_0 and α.
   (*Hint:* Rearrange the equations, after putting in the given values, and form two simultaneous equations for $\dfrac{1}{R_0}$ and α.)

Simultaneous solution of a linear and quadratic equation

There are a few instances when the mathematical model of an engineering problem results in the display of information as two separate equations—one linear and the other a quadratic. It is then necessary to solve them simultaneously in order to obtain an answer to the problem.

The method of solution is to put one of the unknowns in terms of the other using the simple linear equation. We then substitute this into the quadratic equation so that it then has only one unknown—this can then be solved in the usual way.

EXAMPLE 4.24

One item of an oil distillation plant, which our company is manufacturing, is a container (as shown in Fig. 4.6) for storing heavy fuel oil. It has a cylindrical base, whilst the top is conical with an included angle of 120°. We know that its total outer surface area is $107\,\text{m}^2$ and its overall height is $4\,\text{m}$, but the dimensions we need are the diameter of the tank and the height of the cylindrical portion.

As can be seen from Fig. 4.6, we need to find radius r and height H. Now from the right-angled triangle in the root portion we have:

$$\frac{r}{h} = \tan 60° \qquad \text{and} \qquad \frac{r}{l} = \sin 60°$$

giving
$$h = \frac{r}{\tan 60°} \qquad \text{and} \qquad l = \frac{r}{\sin 60°}$$

or
$$h = 0.5774r \qquad \text{and} \qquad l = 1.155r$$

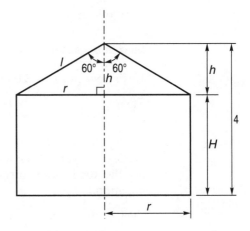

Fig. 4.6

You should notice, that although we shall almost certainly give the final results to 3 s.f., we are working to 4 s.f. accuracy in order to avoid any possible accumulated errors from successive rounding off.

Now since we know that the overall height is 4 m, then

we have $\qquad H + h = 4 \qquad$ and substituting $\quad h = 0.5774r$

then $\qquad H + 0.5774r = 4$

from which $\qquad H = 4 - 0.5774r$

Now: $\left(\begin{array}{c}\text{the total outer} \\ \text{surface area}\end{array}\right) = \left(\begin{array}{c}\text{base} \\ \text{area}\end{array}\right) + \left(\begin{array}{c}\text{cylinder} \\ \text{wall}\end{array}\right) + \left(\begin{array}{c}\text{conical} \\ \text{roof}\end{array}\right)$

$$= \pi r^2 + 2\pi rH + \pi rl$$

Now the surface area $= 107\, m^2$, and substituting $H = 4 - 0.5774r$ and $l = 1.155r$ then we have:

$$107 = \pi r^2 + 2\pi r(4 - 0.5774r) + \pi r(1.155r)$$

and dividing through by π to simplify,

then $\qquad \dfrac{107}{\pi} = r^2 + 2r \times 4 - 2r \times 0.5774r + 1.155r^2$

from which $\qquad 34.06 = r^2 + 8r - 1.155r^2 + 1.155r^2$

or $\qquad r^2 + 8r - 34.06 = 0$

This is a quadratic equation for r, and since it does not factorise easily, we will use the formula

$$r = \frac{-b \pm \sqrt{b^2 - 4ac}}{2a}$$

and here we have $\qquad a = 1, b = 8 \quad$ and $\quad c = -34.06$

Thus $\qquad r = \dfrac{-8 \pm \sqrt{8^2 - 4 \times 1 \times (-34.06)}}{2 \times 1}$

from which $\qquad r = 3.08 \qquad$ or $\qquad -11.08$

Since a negative answer is meaningless here it may be disregarded.

If we now put $r = 3.08$ into the equation for H,

Then $\qquad\qquad\qquad H = 4 - 0.5774 \times 3.08$

$$= 2.22$$

We leave you to check that these values of r and H satisfy the equation for the total outer surface area. Assuming these values satisfy the equation,

then $\quad$ diameter of tank $= 2 \times 3.08 = 6.16$ m

and $\qquad$ cylinder height $= 2.22$ m

Exercise 4.7

Solve simultaneously, by substitution:

1) $y = 3x^2 - 3x - 1$
   $y = 4x - 3$

2) $y = 8x^2 - 2$
   $y = 1 - 5x$

3) $y = 4x^2 - 2x - 1$
   $x - y + 1 = 0$

4) $x = y^2 - 0.4y - 1.5$
   $x + 0.7y + 0.3 = 0$

5) A workshop specialises in the manufacture of precision measuring equipment. A groove is being machined in a block of steel. This groove has a cross-section shaped like an isosceles trapezium which is $2x$ mm wide at the top, $2y$ mm wide at the bottom and has a vertical height of 8 mm. The cross-sectional area of the groove is 144 mm², and its perimeter (two sloping sides and the base) is 32 mm. It can be shown that:

$$x + y = 18$$

and

$$x^2 - 2xy + 32y = 192$$

Solve these equations and hence find the bottom width of the groove cross-section.

Now sketch, or draw to scale, the cross-section so that you can explain how the two solutions to the problem are both valid.

6) Figure 4.7 shows a hot water cylinder with a surface area of 138 m². Show that

$$3\pi r^2 + 2\pi rh = 138$$

and

$$r + h = 10$$

By solving this pair of simultaneous equations find the values of r and h.

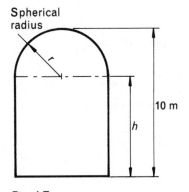

Spherical radius

10 m

h

Fig. 4.7

Transposition of formulae

When solving a linear equation we grouped all the terms containing the unknown, say x, on one side (usually the LHS) and the remainder on the other. We simplified the equation as we went along until a value was obtained for x. This rearrangement could have been called a transposition for x.

However, many equations, or formulae as they are often known, contain several symbols representing unknown quantities, and possibly numbers (called constants) as well.

Such an equation, containing three unknowns but no constants, is $W = IV$. This is a formula in which W is called the subject and is given in terms of I and V. We may wish to transpose this equation for V, which is the same as making V the subject or rearranging for V in terms of I and W.

Maintaining equality of both sides of an equation

As you will see from the worked examples which follow, we use the same rules for maintaining the balance between both sides of an equation as were used previously. The principal object is to group all the terms containing the subject on one side (usually the LHS) and the remainder on the other.

EXAMPLE 4.25

Power W, current I and voltage V, are connected by the formula $W = IV$. Transpose to make V the subject.

Divide both sides by I :
$$\frac{W}{I} = \frac{\cancel{I}V}{\cancel{I}}$$

or
$$\frac{W}{I} = V$$

$$\therefore \quad V = \frac{W}{I}$$

Check: It is possible to check whether the transposition has been made correctly by substituting numerical values.

Putting
$$\left. \begin{array}{l} I = 6 \\ V = 4 \end{array} \right\} \quad \text{into} \quad W = IV$$

we get $\quad W = 6 \times 4 = 24$

Now if we use the transposed form of the given equation,

Putting
$$\left. \begin{array}{l} W = 24 \\ I = 6 \end{array} \right\} \quad \text{into} \quad V = \frac{W}{I}$$

we get $\quad V = \dfrac{24}{6} = 4$

and this verifies the correctness of the transposition.

You may say this was unnecessary as it was obvious the transposition was correct—but this is only because the original equation was comparatively simple. Just accept that this is the procedure and use it for the more difficult formulae.

EXAMPLE 4.26

Heat energy, H, is given by $H = I^2Rt$ where I is current, R is resistance and t is time. Make R the subject of the equation.

Divide both sides by I^2t :
$$\frac{H}{I^2t} = \frac{\cancel{I^2}R\cancel{t}}{\cancel{I^2}\cancel{t}}$$

or
$$\frac{H}{I^2t} = R$$

$$\therefore \quad R = \frac{H}{I^2t}$$

Check: Put $I = 3$, $R = 2$ and $t = 7$ and use the method given in the previous example.

EXAMPLE 4.27

The expression $R = \dfrac{V}{I}$ relates resistance R, voltage V, and current I.

Express V in terms of R and I.

Multiply both sides by I $\qquad\qquad\qquad R \times I = \dfrac{V}{\cancel{I}} \times \cancel{I}$

or $\qquad\qquad\qquad\qquad\qquad\qquad RI = V$

$\therefore \qquad\qquad\qquad\qquad\qquad\qquad V = RI$

Check this result for yourself.

EXAMPLE 4.28

The tension, T, in a cord which is whirling a mass, m, round in a circular

path of radius, r, and tangential velocity, v, is given by $\quad T = \dfrac{mv^2}{r}$.

Transpose for radius r.

Multiply both sides by r $\qquad\qquad\quad T \times r = \dfrac{mv^2}{\cancel{r}} \times \cancel{r}$

$\therefore \qquad\qquad\qquad\qquad\qquad\qquad Tr = mv^2$

Divide both sides by T $\qquad\qquad\qquad \dfrac{\cancel{T}r}{\cancel{T}} = \dfrac{mv^2}{T}$

$\therefore \qquad\qquad\qquad\qquad\qquad\qquad r = \dfrac{mv^2}{T}$

Again you may check this for yourself.

EXAMPLE 4.29

Temperature, t, on the Celsius scale, and temperature, T, on the Absolute scale are related by $\quad T = t + 273$. Make t the subject.

Subtract 273 from both sides $\Big\}$ $\qquad T - 273 = t + \cancel{273} - \cancel{273}$

giving $\qquad\qquad\qquad\qquad T - 273 = t$

or $\qquad\qquad\qquad\qquad\qquad\quad t = T - 273$

Check this result for yourself.

EXAMPLE 4.30

Pressure p, at a depth h, in a fluid of density ρ, is given by $\quad p = p_a + \rho g h$ where p_a is atmospheric pressure and g the gravitational constant. Express h in terms of the other symbols.

Subtract p_a from both sides $\Big\}$ $\qquad p - p_a = \cancel{p_a} + \rho g h - \cancel{p_a}$

giving $\qquad\qquad\qquad\qquad p - p_a = \rho g h$

Divide both sides by ρg $\qquad\qquad \dfrac{p - p_a}{\rho g} = \dfrac{\cancel{\rho g} h}{\cancel{\rho g}}$

or $\qquad\qquad\qquad\qquad\qquad\quad h = \dfrac{p - p_a}{\rho g}$

Check this result for yourself.

EXAMPLE 4.31

F_1 and F_2 are the tensions in the tight and slack sides of a belt drive over a pulley wheel. The velocity, v, of the belt is given by

$$v = \frac{P}{F_1 - F_2}$$ where P is the power transmitted. Make F_2 the subject of this formula.

Multiply both sides $\Big\}$ by $(F_1 - F_2)$

$$v(F_1 - F_2) = \frac{P}{\cancel{(F_1 - F_2)}} \times \cancel{(F_1 - F_2)}$$

giving

$$v(F_1 - F_2) = P$$

Divide both sides $\Big\}$ by v

$$\frac{\cancel{v}(F_1 - F_2)}{\cancel{v}} = \frac{P}{v}$$

giving

$$F_1 - F_2 = \frac{P}{v}$$

Subtract F_1 from $\Big\}$ both sides

$$\cancel{F_1} - F_2 - \cancel{F_1} = \frac{P}{v} - F_1$$

giving

$$-F_2 = \frac{P}{v} - F_1$$

Multiply both sides $\Big\}$ by (-1)

$$-F_2(-1) = \left(\frac{P}{v} - F_1\right)(-1)$$

giving

$$F_2 = \frac{P}{v}(-1) - F_1(-1)$$

$$F_2 = F_1 - \frac{P}{v}$$

Check: Let us use $F_1 = 3$, $F_2 = 1$ and $P = 8$ in the given equation:

$$v = \frac{8}{3 - 1} = 4$$

and putting $F_1 = 3$, $P = 8$ and $v = 4$ into the transposed form, then:

$$F_2 = 3 - \frac{8}{4} = 1$$

and this verifies the transposition.

EXAMPLE 4.32

The resistance, R, of a wire after a temperature rise, t, is given by $R = R_0(1 + \alpha t)$ where R_0 is the resistance at zero temperature and α is the temperature coefficient of resistance. Transpose for t.

We may approach this problem in two ways:

(a) Divide both sides by R_0

$$\frac{R}{R_0} = \frac{\cancel{R_0}(1 + \alpha t)}{\cancel{R_0}}$$

giving

$$\frac{R}{R_0} = 1 + \alpha t$$

Subtract 1 from $\Big\}$ both sides

$$\frac{R}{R_0} - 1 = \cancel{1} + \alpha t - \cancel{1}$$

giving
$$\alpha t = \frac{R}{R_0} - 1$$

Divide both sides by α
$$\frac{\alpha t}{\alpha} = \frac{(R/R_0 - 1)}{\alpha}$$

So the first solution
$$t = \frac{1}{\alpha}\left(\frac{R}{R_0} - 1\right)$$

b) Remove the bracket
$$R = R_0 + R_0 \alpha t$$

Subtract R_0 from both sides $\Big\}$
$$R - R_0 = R_0 \alpha t$$

Divide both sides by $R_0 \alpha$ $\Big\}$
$$\frac{R - R_0}{R_0 \alpha} = \frac{R_0 \alpha t}{R_0 \alpha}$$

Thus the second solution
$$t = \frac{R - R_0}{R_0 \alpha}$$

We will leave you to verify that the two solutions are the same.

Transposition with subject in more than one term

EXAMPLE 4.33

The percentage profit, P, made in a transaction where the selling price is s, and the buying price was b, is given by

$$P = \frac{s - b}{b} \times 100$$

Make b the subject of this equation.

Multiply both sides by b
$$P \times b = \frac{(s - b)}{b} \times 100b$$

and removing the bracket
$$Pb = 100s - 100b$$

Add $100b$ to both sides
$$Pb + 100b = 100s - 100b + 100b$$

giving
$$b(P + 100) = 100s$$

Divide both sides by $(P + 100)$ $\Big\}$
$$\frac{b(P + 100)}{(P + 100)} = \frac{100s}{(P + 100)}$$

giving
$$b = \frac{100s}{(P + 100)}$$

Check this transposition for yourself.

EXAMPLE 4.34

A formula for resistances connected in parallel in an electrical circuit is

$$R = \frac{R_1 R_2}{(R_1 + R_2)} \text{ Transpose for } R_1.$$

The LCM of the RHS is $(R_1 + R_2)$ so we

multiply both sides of the given equation by $(R_1 + R_2)$ $\Big\}$ $R(R_1 + R_2) = \dfrac{R_1 R_2}{\cancel{(R_1 + R_2)}} \times \cancel{(R_1 + R_2)}$

and remove brackets $RR_1 + RR_2 = R_1 R_2$

and taking RR_1 from both sides $\Big\}$ $\cancel{RR_1} + RR_2 - \cancel{RR_1} = R_1 R_2 - RR_1$

from which $R_1 R_2 - RR_1 = RR_2$

and factorise the LHS $R_1(R_2 - R) = RR_2$

Divide through by $(R_2 - R)$ $\dfrac{R_1 \cancel{(R_2 - R)}}{\cancel{(R_2 - R)}} = \dfrac{RR_2}{(R_2 - R)}$

thus $R_1 = \dfrac{RR_2}{(R_2 - R)}$

We leave you to verify the result.

Exercise 4.8

1) In the thermodynamics the characteristic gas equation is $pV = nRT$ which connects pressure, p, volume, V, and temperature, T. Make T the subject of this formula.

2) A large cone has radius, R, and height, H. A small cone has radius, r, and height, h. These dimensions are related by $\dfrac{R}{r} = \dfrac{H}{h}$ Rearrange this equation for h.

3) The equation of motion $v = u + at$ is for movement with constant acceleration, a. The initial velocity is u, and the final velocity is v, arrived at after time, t. Make u the subject of the equation.

4) Using $v = u + at$ again, transpose for t.

5) Temperature is degrees Fahrenheit F and in degrees Celsius C are related by $F = \dfrac{9}{5}C + 32$. Rearrange this formula for C.

6) The standard equation of a straight line is $y = mx + c$ in co-ordinate geometry. Rearrange to make x the subject of the equation.

7) The sum to infinity, S, of a geometric progression is given by the expression $S = \dfrac{a}{1 - r}$ where a is the first term and r is the common ratio. Transpose to make r the subject.

8) The current, I, and the voltage, V, in a circuit containing two resistances R and r in series are connected by the formula $I = \dfrac{V}{R+r}$. Transpose for R.

9) The surface area, S, of a cone having height h and base radius r is given by $S = \pi r(r+h)$. Find h in terms of π, r and S.

10) The expression $H = ws(T-t)$ is used in finding total heat, H. Rearrange to make T the subject of the formula.

11) The expression $\dfrac{v^2 - u^2}{2a} = s$ is a law of motion for constant acceleration, a, the distance travelled, s, and the initial and final velocities, u and v, respectively. Transpose the equation for u^2.

12) The common difference, d, of an arithmetic progression whose sum S to n terms is given by $d = \dfrac{2(S-an)}{n(n-1)}$ where the first term is a. Transpose this formula: a) for S b) for a

13) The expression $R = \dfrac{1}{1/R_1 + 1/R_2}$ is for resistances in parallel in an electrical circuit. Transpose for R_1.

14) Another form of the expression for parallel resistances in an electrical circuit is $R = \dfrac{R_1 R_2}{R_1 + R_2}$. Make R_2 the subject of this equation.

Transposition of formulae containing roots and powers

The same basic principles of transposition apply here. In Example 4.35 the new subject, albeit squared, has been isolated on the LHS—thus to obtain u we take the square root of both sides, so maintaining the balance of the equation.

In Example 4.36 the new subject is 'inside the square root'. If we square both sides of the equation we can get rid of the square root and then carry on the transposition as usual.

EXAMPLE 4.35
An equation of motion for constant acceleration, a, is $v^2 = u^2 + 2as$ where the initial velocity, u, is changed to the final velocity, v, after travelling a distance s. Make u the subject of the equation.

Subtract $2as$ from both sides $v^2 - 2as = u^2 + \cancel{2as} - \cancel{2as}$

giving $u^2 = v^2 - 2as$

and taking the square root of both sides $\Big\}$ $\sqrt{u^2} = \sqrt{v^2 - 2as}$

giving $u = \sqrt{v^2 - 2as}$

Check this result for yourself.

EXAMPLE 4.36

The period, t, of a simple pendulum having length l is given by

$$t = 2\pi\sqrt{\frac{l}{g}}$$ where g is the gravitational constant. Transpose this

equation for l.

Divide both sides by 2π $$\frac{t}{2\pi} = \sqrt{\frac{l}{g}}$$

Square both sides $$\left(\frac{t}{2\pi}\right)^2 = \left(\sqrt{\frac{l}{g}}\right)^2$$

or $$\frac{t^2}{4\pi^2} = \frac{l}{g}$$

Multiply both sides by g $$l = \frac{t^2 g}{4\pi^2}$$

Again we leave the checking to you.

Exercise 4.9

1) The velocity, v, of a jet of water is given by the expression $v = \sqrt{2gh}$. Transpose this for head of water, h.

2) The formula $A = \pi r^2$ gives the area, A, of a circle in terms of its radius, r. Find r in terms of π and A.

3) The kinetic energy, E, of a mass m travelling at velocity v is given by $E = \frac{1}{2}mv^2$. Make v the subject of this expression.

4) The diameter, d, of a circle of area A is given by $d = 2\sqrt{\frac{A}{\pi}}$. Find A in terms of π and d.

5) The strain energy, U, of a material under stress is given by $U = \frac{f^2 V}{2E}$. Rearrange this for f.

6) In a right-angled triangle, $b = \sqrt{a^2 - c^2}$. Transpose for c.

7) The frequency, f, of a simple pendulum is given by $f = \frac{1}{2\pi}\sqrt{\frac{g}{l}}$. Make l the subject of this expression.

8) A shaft is acted on by a bending moment, M, and a torque, T. The equivalent torque, T_e, is given by $T_e^2 = M^2 + T^2$. Transpose this for M.

9) The crippling load on a strut, according to the Euler theory, is given by $P = \frac{\pi^2 EI}{(CL)^2}$. Make C the subject of this equation.

10) The total energy, E_t, of a mass m at a height h above a given datum and travelling with a velocity v is given by the equation $E_t = mgh + \frac{1}{2}mv^2$. Find an expression for velocity v.

11) A property of a solid rectangular block, called the radius of gyration, k, is given by $k = \sqrt{\dfrac{a^2 + b^2}{12}}$. Rearrange this equation for b.

12) A property of a solid cylinder, called the radius of gyration, k, is given by $k = \sqrt{\dfrac{L^2}{12} + \dfrac{R^2}{4}}$. Make L the subject of this equation.

13) A formula connected with stress in cylinders is $\dfrac{D}{d} = \sqrt{\dfrac{f + p}{f - p}}$. Transpose this for f.

14) A formula for the equivalent shear load in the design of a bolt is $Q_e = \frac{1}{2}\sqrt{P^2 + Q^2}$. Rearrange this to make an equation for P.

15) A formula for the equivalent tensile load, P_e, in the design of a bolt, is $P_e = \frac{1}{2}(P + \sqrt{P^2 + 4Q^2})$. Transpose for Q.

Portfolio problems 1

1) The marketing department of a company selling petroleum products is introducing a brand new product. As an introductory offer they want to sell 1.25 litres for the price of 1 litre. The production manager needs to investigate some of the factors which the marketing department have failed to appreciate, and then forward his report to them. You have been asked to calculate the following:
   a) The amount of extra material, in square metres, required to make a trial batch of 150 000 cans, allowing an extra 10% for joints.
   b) The dimensions of the new trial can.
   c) The mass of the new can when empty, and the percentage increase from the mass of the old 1 litre cans.

   To help you have been given a small design specification:

   1. The height of the trial cans must be the same as the height of the 1 litre cans, namely 200 mm.
   2. The can material has a mass of 1.95 kg per square metre of sheet material used.
   3. The accuracy of your answers should be to three significant figures.

2) A.H. Machine Parts Ltd is a medium sized company manufacturing gib strips, which are used on many machines, e.g. on lathe and milling machine slideways, to remove any play in these slideways. One of them is shown in Fig. P1.1.

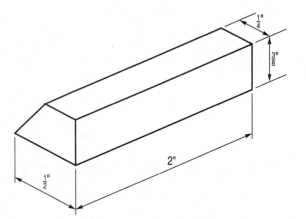

Fig. P1.1

The company is in the process of changing to the metric system. As junior designer you have been asked to metricate this existing component. There are now, however, several knock-on problems. The shipping specifications will have to be amended. An order for 2500 of these components has already been taken and your amendment to the design and shipping statistics is needed urgently.

Using Fig. P1.1, give details of the metric dimensions and mass of each component, and the mass of the total shipment. Take the density of the material as 7900 kg m^{-3}. Justify any assumptions you need to make.

3) You are responsible for the production of 1000 spools, as shown in Fig. P1.2, per week in a machine shop. These spools are required for the manufacture of pneumatic valves.

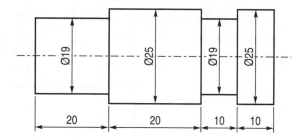

Fig. P1.2

You have been asked by the financial manager to improve profit margins, and believe that one simple way would be to sell the scrap metal (swarf), which is currently thrown away. The components are produced from round steel bars 30 mm diameter, the steel having a density of 7800 kg m^{-3}. Apart from the loss of metal when turning down to the required diameters there is a further waste, equivalent to 5 mm of the original bar, when parting off each component. How much money could you make in a 48 working week year by selling the scrap at a price of £40 per ton (imperial ton – this is often used by small established scrap companies)? Times are hard and the financial manager is eagerly awaiting your report!

4) The publicity department of a company which manufactures electrical wiring has requested that you should calculate the mass of the 100 metre reels of single core plastic coated copper wire. The 5 mm diameter single solid core is drawn from copper which has a density of 8800 kg m^{-3}. The plastic coating is of a uniform thickness of 2 mm all round the copper core, the density of the plastic being 900 kg m^{-3}. The mass of the reel on which the wire is wound is 0.5 kg. Find and state the mass to an accuracy you consider suitable for the information leaflet where data will be given.

5) You work for a company which produces motorway crash barriers. The department of road transport would like your company to design a new barrier, resulting from the increase in the maximum mass of a vehicle from 36 tonnes to 40 tonnes. You need to calculate the force with which a 40 tonne truck would hit the barrier if the velocity prior to collision was 100 km/h reducing to zero in 0.5 seconds.

Use the formulae:

$$\text{force (N)} = \text{mass (kg)} \times \text{acceleration (m s}^{-2})$$

and $\quad$ $$\text{acceleration (m s}^{-2}) = \frac{\text{change in velocity (m s}^{-1})}{\text{time (s)}}$$

and give your answer in preferred units correct to 3 s.f.

6) The plate shown in Fig. P1.3 is to be used to join two plates together. The chief inspector has asked you to design a template so that the inspectors will be able to check these components. As shown, two holes, of diameters D_1 and D_2, respectively, are drilled in the steel plate. The distance C between the hole centres is to be 40.0 ± 0.02, the diameter D_1 is 7.5 ± 0.01, and the diameter D_2 is 12.5 ± 0.01, all dimensions being in millimetres.

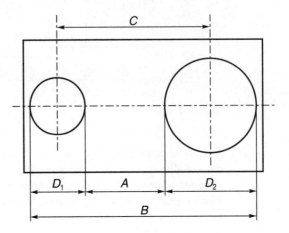

Fig. P1.3

You will have to find the maximum and minimum values of the dimensions A and B. What are these values?

7) You are employed by a lift manufacturer who has been asked to install a lift in a new office block. A colleague has been to the customer and established the specifications for the lift. He has designed the lift shaft and the lift cage and now you need to calculate the diameter of the wire which will support the cage.

Specification
The lift is to carry up to four adults, assuming the maximum mass per adult to be 130 kg. The cage of the lift when assembled will have a mass of $\frac{1}{2}$ tonne. The value of Young's modulus (E) for the steel wire $= 207\,\text{GN/m}^2$ and your assistant has calculated the strain acceptable in the steel to be 2.5×10^{-5}.

The formulae required for the calculation were obtained from the engineers' handbook at your company and are shown below:

$$\text{Young's modulus } E = \frac{\text{Stress }(\sigma)}{\text{Strain }(\varepsilon)} \qquad\qquad \text{Area of circle} = \frac{\pi D^2}{4}$$

$$\text{Stress } \sigma = \frac{\text{Force }(F)}{\text{Area }(A)}$$

Calculate the diameter of the wire, showing your calculations clearly.

8) You are a test engineer for a company who make moving coil instruments. The design department has just completed a prototype ammeter and would like you to test the shunts that are available on the meter. You need to calculate the f.s.d. value shown on the meter when each of the shunts is used. The designers have given you the values of the current and resistance for the meter and the two shunts below.

The arrangement is as shown in Fig. P1.4.

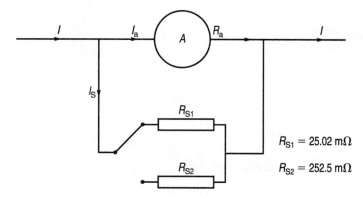

$R_{S1} = 25.02\ m\Omega$

$R_{S2} = 252.5\ m\Omega$

Fig. P1.4

Without the shunts the ammeter has a full scale defection (f.s.d.) when the current is 50 mA and it has a resistance of 25 Ω.

By adding a shunt the ammeter is able to measure higher current values, as the shunt takes up the additional current. The designers have enabled two shunts to be used on the meter by the flip of a switch.

You are given the formulae:

$$R_s = \frac{I_a R_a}{I_s} \qquad I_s = I - I_a$$

where
R_s = the resistance in the shunt
R_a = the resistance of the meter
I = the total circuit current at f.s.d.
I_s = the current flowing in the shunt
I_a = the current flowing in the meter/instrument

Calculate the current value at f.s.d. when each of the shunts is in use.

9) A design project on which you are working involves the investigation of the linear dimensions, the surface area and the volume of a watertight conical float, which is fabricated from thin copper sheet. Now, for a right circular cone, as shown in Fig. P1.5, the following formulae apply:

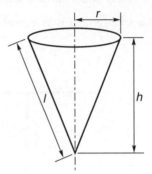

Fig. P1.5

Curved surface area $= \pi r l$

$$\text{Volume} = \tfrac{1}{3}\pi r^2 h$$

The only information you can find about the float is that the slant length $l = 45$ mm, and when hung from a sensitive spring balance it weighed 0.196 N. The information required by your design colleagues is the value of the radius, r, the height, h, and the volume, in litres, of water displaced when the float is fully immersed. Quite reasonably you have decided to ignore the effect of any joints etc. As you puzzle out how to begin the technician who fabricated the cone volunteers that the material used was sheet copper which has a mass of 4.8 kg per square metre of sheet. Now you should be able to go ahead.

10) A customer from the chemical industry asks you to find the volume of a cylindrical can, as shown in Fig. P1.6, which will be used as a pressurised container for a fly spray. The can is unusual insofar as it has hemispherical ends, one convex and the other concave. Apart from the height of the container being 134 mm, the only other information available is that the outer surface area of the metal used in its construction, neglecting any joint effects etc., is 0.0253 square metres. Luckily a 'whizz kid' friend of yours took one glance and said, with much sarcasm, that at least one thing was obvious from the diagram, namely that $r + l = 134$ and then disappeared. Now it is time to make use of the other piece of information and form another equation involving r and l. You will then have two equations which you can solve simultaneously and so find the volume for your customer.

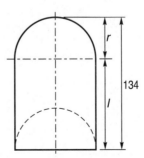

Fig. P1.6

Portfolio assignment 1

In the four chapters covered in this part of the book, we have looked at the use of number and algebra. The key mathematical methods and laws have been introduced and, hopefully, you will have worked through some, if not all, of the portfolio problems which show how the theories are applied in real situations. The next step is to tackle more extensive problems which cover the whole range of the GNVQ element, and can also be used as evidence in your own portfolio. An example assignment for number and algebra is shown below.

The crankshaft

Design brief

A motorcycle manufacturer is developing a new machine, and wishes to increase the engine capacity without increasing the overall weight. This could be done by increasing the throw of the crankshaft, whilst being careful not to increase its volume.

The chief design engineer at your company has given you the drawing, Fig. P1.7, for the current design of crankshaft. He suggests that you increase the throw from 50 mm to 60 mm and, by using a steel of greater strength, make the 25 mm diameters smaller so that the overall volume of the crankshaft remains unaltered.

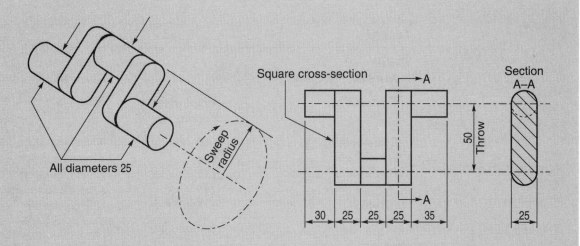

Fig. P1.7

(Engine experts: please do not scoff at this drawing of a crankshaft – it is for calculation purposes only!)

The production and manufacturing sections of the company have completed the planning details for producing the crankshafts and the sales team are negotiating a contract to produce 100,000 per year for the next four years.

Table P1 shows the costs and times of manufacture. The financial director has asked for an estimate of how much profit the company will make on this contract.

Task 1

a) Formulate an equation from the area of section A–A in terms of the diameter of the crankshaft.
b) Calculate the area of the section A–A.
c) Using your answers to parts a) and b), formulate a quadratic equation to find the diameter of the new crankshaft, if the throw is increased to 60 mm and the area of the new section kept the same as section A–A.
d) Calculate the new crankshaft sweep radius.

Task 2

a) Formulate an equation for the cost of producing x number of crankshafts.
b) Formulate an equation for the proceeds of selling x number of crankshafts.
c) Using your answers to parts a) and b), find the number of crankshafts which would need to be sold before the company starts to make a profit.
d) How many crankshafts would need to be sold to make a profit of £10 000?
e) How much profit/loss would there be in the production of 10 000 crankshafts?

Table P1

Variable costs		
Raw materials/power		£20/crankshaft
Operation	**Production cost/hr**	**Time**
Drop forge	£5	10 min/crankshaft
Machining 1	£15	15 min/crankshaft
Machining 2	£15	10 min/crankshaft
Heat treatment	£10	3 hr/10 crankshafts
Grinding	£15	1 hr/crankshaft
Packaging	£5	10 min/crankshaft
Fixed costs		
Capital expenditure (cost of new machinery)	£100 000	
Overheads (e.g. rent, rates, administration)	£300/100 crankshafts	
Transport	£200/200 crankshafts	
Sales		
Selling price	£100 per crankshaft plus 17.5% VAT	

Unit coverage

Table E1 below shows which parts of the unit have been covered in this assignment and where there are opportunities to assess core skills. If mapping of this type is shown for all your portfolio work, it is easy to check that you have covered all the elements in the unit.

Table E1

Task	Unit	Element/PC	Core skills
1	8	1.1, 1.2, 1.3, 1.4	Application of number 3.2.1 and 3.2.6
2	8	1.1, 1.2, 1.3, 1.4	Application of number 3.1.2 and 3.2.3
	1	2.4, 3.3, 3.4, 3.5	Application of number 3.2.1 and 3.2.2

PART TWO

Use of trigonometry to solve engineering problems

The branches of engineering with which we are concerned are involved with the design and manufacture of components and final articles. During these processes much of the detailed work uses sketches, drawings and graphical work of a varied nature. All these diagrams introduce dimensions such as lengths, widths and heights etc. Thus the need to be able to 'solve' basic figures such as triangles, into which more complicated shapes may be divided, requires a little more than the use of a rule and the theorem of Pythagoras. So the use of the right-angled triangle was extended and the basic trigonometric ratio definitions of sine, cosine and tangent were developed.

These ideas were soon introduced into technological applications such as angles of elevation and heights of transmission aerials, frameworks and lengths of drive belts. Workshop applications enabled the use of precision tools for fine measurements.

Once the basic ideas were extended into triangles and circles, we had established a firm base for dealing with more complicated forms.

The ideas of vectors, representing forces and velocities, and of phasors, representing currents and voltages, give rise mostly to triangles, all of which may be solved using trigonometry.

Miscellaneous topics in trigonometry

Cartesian and polar coordinates represent two ways that points may be plotted and thus designated on a grid. This is necessary, for instance, when data has to be prepared for a CNC machine tool.

Inverse trigonometric functions arise from the reverse of the sequence of finding a ratio (sine, cosine or tangent) when given an angle. Surely this is easy? Simply enter the ratio figure and press the correct keys on the calculator and the angle will appear. Yes, but the problem is that there are many angles which have the same ratio—the machine gives us one, but it is our job to be able to sort out any other relevant values.

Finally, we look at the circle. Trigonometry is useful here since chords, diameters, radii and tangents create triangles, also included are some geometric properties useful in many engineering applications, such as fine measurements.

Cartesian and polar representation

Location of a point

Engineers are often concerned about positioning, or locating, a point on a plane. An example of such a plane is the paper on which you are writing, or the flat surface table of a CNC drilling machine.

You are probably aware of Ordnance Survey maps which have horizontal and vertical grid lines. These are suitably numbered and enable exact positioning by giving a grid reference. One of our positioning systems is similar, our grid reference being called Cartesian coordinates.

Another system is by using a radial distance along a line which is at a specified angle from a reference datum line (usually horizontal). These two values are called Polar coordinates.

We will now examine how these two systems—Cartesian and Polar coordinates—are used in practice.

Cartesian axes of reference

We take two lines at right angles to each other, Fig. 5.1, which are called axes of reference. Generally the horizontal line is known as the *x*-axis, the vertical line as the *y*-axis and the point of intersection as **the origin**, O.

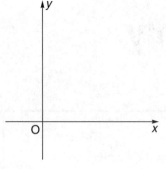

Fig. 5.1

Cartesian coordinates

These are also commonly known as **rectangular** coordinates. More often than not we simply call them coordinates since they are used for the majority of work for positioning and graphs.

Coordinates are used to mark the position of a point.

The *x*-coordinate, measured horizontally, uses the scale on the *x*-axis. The *y*-coordinate, measured vertically, uses the scale on the *y*-axis.

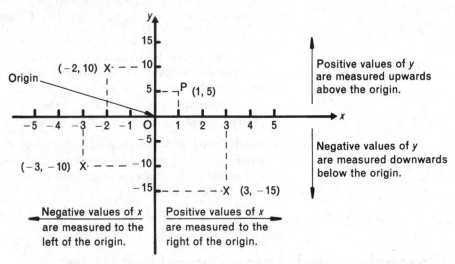

Fig. 5.2

In Fig. 5.2, point P has coordinates $x = 1$ and $y = 5$ and is often known as the point (1, 5). Negative values are catered for by scales similar to those used in directed numbers. Three other points are shown, namely (−2, 10), (−3, −10) and (3, −15).

The distance between two points

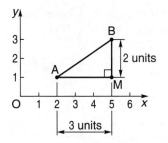

Fig. 5.3

Points A (2, 1) and B (5, 3) are shown in Fig. 5.3. If we complete the triangle ABM we can see that the length of AM is 3 units, and BM is 2 units.

Now AB may be found by the use of the theorem of Pythagoras:

$$AB = \sqrt{3^2 + 2^2}$$
$$= 3.61 \quad \text{correct to 3 s.f.}$$

Strictly there are two square roots of 13, namely +3.61 and −3.61. However, we are only interested in the magnitude of AB—called the modulus of AB.

$$\therefore \quad \text{length AB} = \text{modulus of } \sqrt{3^2 + 2^2} \quad \text{or} \quad |\sqrt{3^2 + 2^2}|$$
$$= +3.61$$

Thus the distance between holes A and B is 3.61 correct to 3 s.f.

EXAMPLE 5.1
Find the distance between the hole centres on the circular steel plate shown in Fig. 5.4.

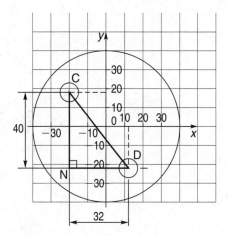

Fig. 5.4

Somebody has already decided (quite sensibly it would apear) to set the reference axes as shown, and designate hole C as (−20, 18) and hole D as (12, −22).

We again draw a suitable right-angled triangle CND, with sides of length 40 and 32. Using the theorem of Pythagoras:

$$CD = \sqrt{40^2 + 32^2}$$
$$= 51.2$$

You should note here that although negative values are present, these only locate the holes (as coordinates). But we are only concerned about the positive lengths of the lines (their modulii—plural of modulus).

Thus the distance between hole centres is 51.2 mm correct to 3 s.f.

Polar coordinates

You have met previously Cartesian (or rectangular) coordinates with which P may be given as the point (x, y) as shown in Fig. 5.5.

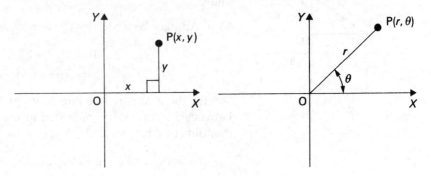

Fig. 5.5   Fig. 5.6

Another way of giving the location of P is by using its distance r from the origin O, together with angle θ that OP makes with the horizontal axis OX (Fig. 5.6). Figure 5.7 shows five typical points, plotted on a polar grid, together with their respective polar coordinates.

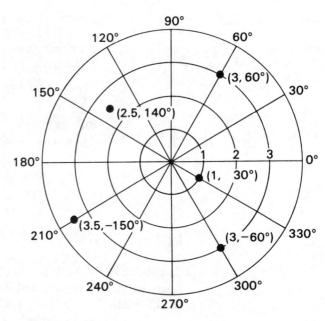

Fig. 5.7

Positive values of θ are always measured anticlockwise from OX whilst negative values are measured clockwise (see Fig. 5.7).

Polar graph paper is available but is not so readily obtained as the common linear variety. However, points with polar coordinates may be plotted without any difficulty on a rectangular axis grid.

A typical practical application of a polar plot would be for values of luminous intensity of an electric light bulb, to show how it varied around the bulb relative to the position of the filament etc.

Relationship between cartesian and polar coordinates

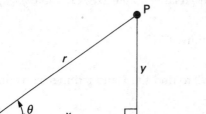

Fig. 5.8

From the right-angled triangle POM in Fig. 5.8 we can see that:

$$x = r \cos \theta$$

and

$$y = r \sin \theta$$

Also

$$r = \sqrt{x^2 + y^2}$$

and

$$\tan \theta = \frac{y}{x}$$

Using the above relationships it is reasonably easy to convert from Cartesian to polar coordinates, and vice versa. Always make a sketch of the problem because this will enable you to see which quadrant you are dealing with.

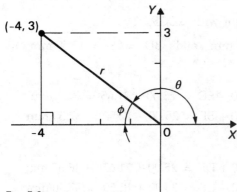

Fig. 5.9

EXAMPLE 5.2

Find the polar coordinates of the point $(-4, 3)$.

From Fig. 5.9

$$\tan \phi = \frac{3}{4} = 0.75$$

$$\therefore \qquad \phi = 36.9°$$

Thus

$$\theta = 180° - 36.9° = 143.1°$$

Also

$$r = \sqrt{3^2 + 4^2} = 5$$

Thus the point is $(5, 143.1°)$ in polar form.

EXAMPLE 5.3

Express in Cartesian form the point $(6, -129°)$

From Fig. 5.10

$$\phi = 180° - 129° = 51°$$

Thus

$$x = 6 \cos 51° = 3.78$$

and

$$y = 6 \sin 51° = 4.66$$

Now, having drawn a diagram, we can see that both the x and y coordinates are negative.

Thus the Cartesian form of the point is $(-3.78, -4.66)$.

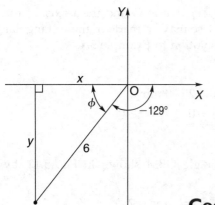

Fig. 5.10

Coordinate hole dimensions

In marking-out and in operating certain machine tools it is convenient to give the dimensions of holes relative to two axes which are at right angles to each other.

In graphical work the position of a point on a graph is specified by its coordinates (that is, the distances at which the point lies from the x- and y-axes, respectively). Coordinate hole centres are specified in exactly the same way.

EXAMPLE 5.4

Three holes are to have their centres equally spaced on a 50.00 mm pitch circle diameter as shown in Fig. 5.11. Calculate the coordinate dimensions of the holes, relative to the axes Ox and Oy.

Hole A

x dimension $= 25.00$ mm
y dimension $= 0$

For the holes B and C, to find the x and y dimensions draw the $\triangle$DCF and the $\triangle$BED.

In $\triangle$BED, $\dfrac{ED}{BD} = \cos 30°$

$\therefore$ $ED = BD(\cos 30°) = 25(\cos 30°) = 21.65$ mm

and $\dfrac{BE}{BD} = \sin 30°$

$\therefore$ $BE = BD(\sin 30°) = 25(\sin 30°) = 12.50$ mm

Since $\triangle$BED is congruent with $\triangle$DCF,

$ED = DF = 21.65$ mm   and $BD = CF = 12.50$ mm

Hole B

x dimension $= 25.00 - ED = 25.00 - 21.65 = 3.35$ mm
y dimension $= 25.00 + BE = 25.00 + 12.50 = 37.50$ mm

Hole C

x dimension $= 25.00 + DF = 25.00 + 21.65 = 46.65$ mm
y dimension $= 25.00 + CF = 25.00 + 12.50 = 37.50$ mm

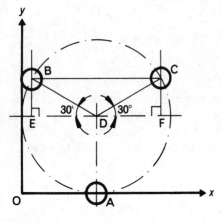

Fig. 5.11

EXAMPLE 5.5

Find the polar coordinates, relative to O and the x-axis, of the holes A and B in Fig. 5.11. Use may be made of the rectangular coordinates obtained as a solution to Example 5.4.

Hole A

$$OA = 25.00 \text{ mm}$$
$$\theta = 0°$$

Hole B

Using the right-angled triangle OBM shown in Fig. 5.12, by Pythagoras we have:

$$OB^2 = 3.35^2 + 37.50^2$$

Thus $OB = 37.65$ mm

Also $\tan \theta = \dfrac{37.50}{3.35} = 11.194$

giving $\theta = \text{inv tan } 11.194$
$$= 84.9°$$

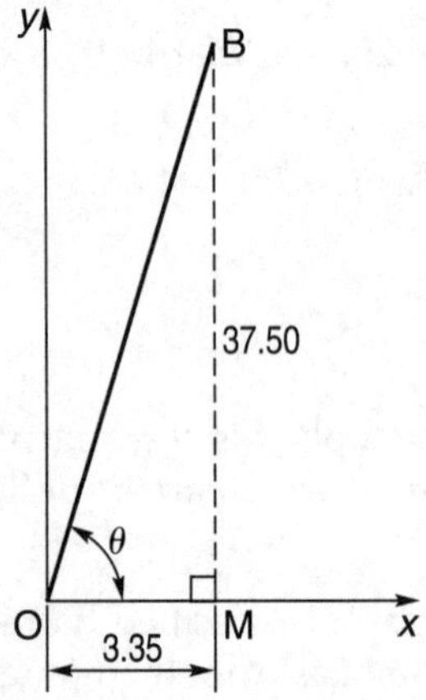

Thus the polar coordinates of holes A and B are (25.00, 0°) and (37.65, 84.9°), respectively.

Fig. 5.12

Exercise 5.1

1) Fig. 5.13 shows 5 equally spaced holes on a 100 mm pitch circle diameter. Calculate their coordinate dimensions relative to the axes O*x* and O*y*.

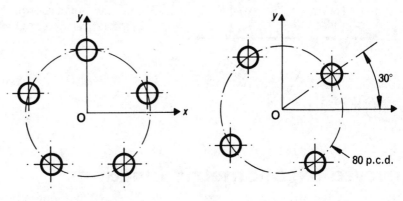

Fig. 5.13   Fig. 5.14

2) **Calculate the polar coordinates of the holes, relative to O and the *x*-axis, of the holes in Fig. 5.13 using the details in question 1.**

3) **Four holes are equally spaced as shown in Fig. 5.14. Find their rectangular coordinate dimensions, relative to the axis O*x* and O*y*.**

4) **Calculate the polar coordinates of the holes shown in Fig. 5.14 relative to O and the *x*-axis.**

5) **Find the Cartesian dimensions for the three holes shown in Fig. 5.15 relative to the axis O*x* and O*y*.**

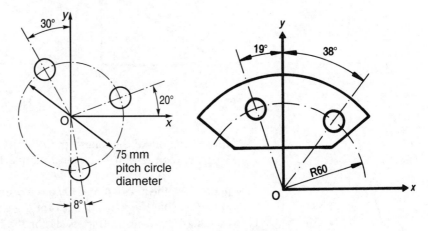

Fig. 5.15   Fig. 5.16

6) Find the polar coordinates for the three holes in Fig. 5.15 relative to O and the axis O*x*.

7) Find the rectangular coordinates for the centres of the two holes shown in Fig. 5.16, relative to the *x*-axis and *y*-axis.

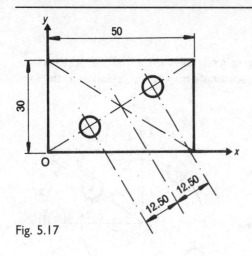

Fig. 5.17

8) Find the polar coordinates for the two holes, relative to O and axis Ox, shown in Fig. 5.16.

9) Find the Cartesian dimensions for the two holes shown in Fig. 5.17, relative to the axis Ox and Oy.

10) Find the polar coordinates for the two holes shown in Fig. 5.17, relative to O and the x-axis.

Inverse trigonometric functions

The method we will be using when determining the values of the angles involves the use of the three principal trigonometric waveforms, namely the sine, cosine and tangent curves. So you will start by finding out all about these graphs—a thorough knowledge here is essential; even more so when dealing with electrical waveforms.

Construction of sine and cosine curves

From right-angled triangle OPM in Fig. 5.18 we have $\sin \theta = \dfrac{PM}{OP}$

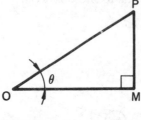

Fig. 5.18

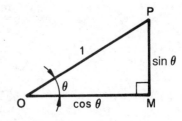

Fig. 5.19

Now if we make the length of OP unity, i.e. OP = 1 unit, as shown in Fig. 5.19, then length PM = $\sin \theta$. Similarly length OM = $\cos \theta$.

In Fig. 5.20, the axes Ox and Oy have been drawn at right angles to each other, just like the x and y axes for a graph. This enables us to use the same sign convention (Fig. 5.21) for the horizontal and vertical scales.

If we now draw a circle, wth centre O and a radius of unity (Fig. 5.22), we can easily fit in $\triangle$OPM for whatever value of angle θ we choose. All angles are measured from Ox as a datum (i.e. starting position). The angles we are familiar with are positive, and measured in an anti-clockwise direction. Negative angles are not so well-known, and are measured clockwise from Ox, as shown in Fig. 5.23.

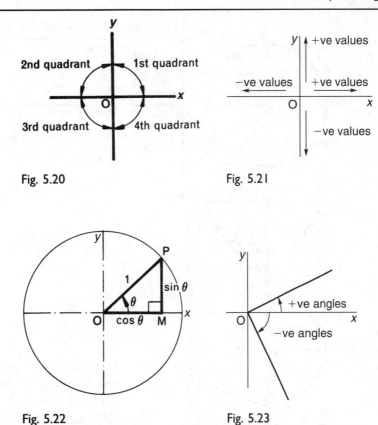

Fig. 5.20

Fig. 5.21

Fig. 5.22

Fig. 5.23

We are now in a position to use these ideas to help with plotting the sine and cosine graphs. If you look at the top portion of Fig. 5.24, you will see the general arrangement.

The cosine curve may also be constructed as shown in Fig. 5.24. However, it is usually drawn with the angle base horizontal (Fig. 5.26), in order that it may be compared with other trigonometrical curves.

It is a good idea for you to actually draw a circle with axes, and the graph scales. A radius of 20 mm and a horizontal scale of 80 mm will fit your notepaper—it doesn't matter if it is not exactly to scale, but you will be surprised how good the result can be! The horizontal angle scale will be from 0 to 360° representing one revolution of the radius OP. The sine curve is obtained by joining up the points where the horizontal projections from P cut the corresponding angle verticals.

You are now able to find the sine value for any particular angle, by simply measuring the vertical height (called the ordinate on a graph). This will give you a direct result only if you started with a circle of radius 1. But if you started with a 20 mm radius circle, and the ordinate measurement is 8.3 mm, then the correct value for the sine of the angle is $\frac{8.3}{20}$ or 0.415. Try this for some values and compare the results with those obtained from your calculator.

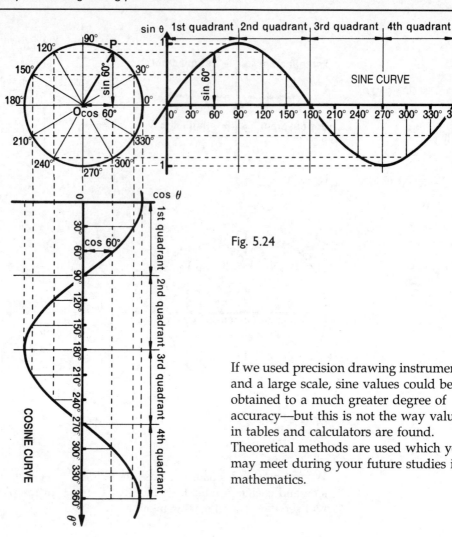

Fig. 5.24

If we used precision drawing instruments and a large scale, sine values could be obtained to a much greater degree of accuracy—but this is not the way values in tables and calculators are found. Theoretical methods are used which you may meet during your future studies in mathematics.

A never-ending curve

The curve we have drawn is for one anti-clockwise revolution of OP and angles 0°–360°. But we could do a second revolution giving 360°–720° and for each revolution of OP we would get a similarly shaped waveform. In fact, the shape repeats itself and the curve is called **cyclic**. Now if OP rotated clockwise, the angle values would be negative if also measured clockwise from O*x*—one revolution would be for angles 0° to −360°. Another revolution would be for −360° to −720° and so on. Again the curve shape would be similar.

The waveform is never-ending, being of infinite length in each direction. We should be careful to show we understand this every time we draw, or even sketch, the waveform by putting 'tails' on each end of our, say, 0°–360° plot, and not finishing the curve exactly at 0° and 360°.

Details of the sine curve

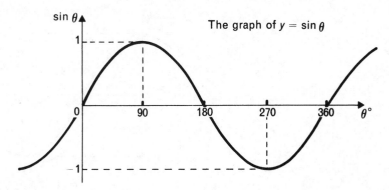

Fig. 5.25

Most of the time, as engineers, we are only interested in the sine curve for angle values 0°–360° which is shown in Fig. 5.25. But it is vital that you have full knowledge of this: you should be able to sketch (not draw accurately as it would take too long) the figure with every detail exactly as it is shown here. You should also know the following facts:

Properties of the sine curve

1) The curve cuts the horizontal axis at 0°, 180° and 360°.

    Thus $\qquad \sin 0° = 0, \quad \sin 180° = 0 \quad$ and $\quad \sin 360° = 0$

2) The maximum value of $\sin \theta$ is $+1$ at 90°.

    Thus $\qquad\qquad\qquad \sin 90° = 1$

3) The minimum value of $\sin \theta$ is -1 at 270°.

    Thus $\qquad\qquad\qquad \sin 270° = -1$

4) Values of $\sin \theta$ are:

           $+$ ve for angles 0°–180°,

           $-$ ve for angles 180°–360°.

Details of the cosine curve

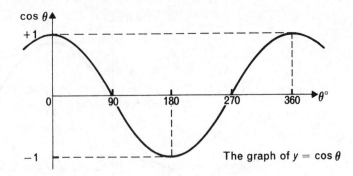

Fig. 5.26

The cosine curve, Fig. 5.26, is similar to the sine curve in all aspects except one. The waveform has been 'shifted bodily horizontally' through 90°. We say that there is a phase difference of 90°. Again it is essential that you can sketch the arrangement and know the following facts:

Properties of the cosine curve

1) The curve cuts the horizontal axis at 90° and 270°.

   Thus $\cos 90° = 0$ and $\cos 270° = 0$

2) The maximum value of $\cos \theta$ is +1 at 0° and 360°.

   Thus $\cos 0° = 1$ and $\cos 360° = 1$

3) The minimum value of $\cos \theta$ is −1 at 180°.

   Thus $\cos 180° = -1$

4) Values of $\cos \theta$ are:

   + ve for angles 0°–90° and 270°–360°,

   − ve for angles 90°–270°.

Details of the tangent curve

Since this is your first meeting with the tangent curve it is a good idea to plot the graph of $y = \tan \theta$. Values for plotting may be obtained using the identity $\tan \theta = \dfrac{\sin \theta}{\cos \theta}$ and the numerical values of sine and cosine from the curves you have drawn previously. Alternatively the values of tangent may be found directly from your calculator.

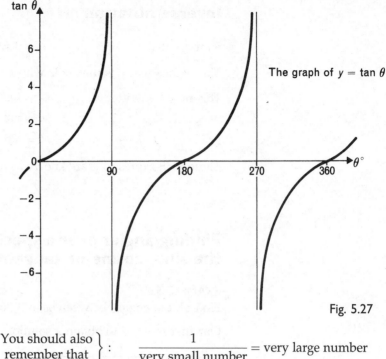

The graph of $y = \tan \theta$

Fig. 5.27

You should also remember that } :
$$\frac{1}{\text{very small number}} = \text{very large number}$$

Thus
$$\frac{1}{\text{zero}} = \text{infinity (symbol } \infty)$$

Although 'disjointed', this is a **cyclic** waveform and, like the sine and cosine waveforms, extends indefinitely in either direction for +ve and −ve angle values.

Although not used as much as the sine and cosine curves, you should still be able to sketch the arrangement in Fig. 5.27 and know the following important facts:

Properties of the tangent curve

1) The curve cuts the horizontal axis at 0°, 180° and 360°.

Thus $\quad\quad \tan 0° = 0, \quad \tan 180° = 0 \quad$ and $\quad \tan 360° = 0$

2) At the discontinuities
$$\tan 90° = \infty \quad \text{and} \quad \tan 270° = \infty$$

3) Values of $\tan \theta$ are:

$\quad\quad$ + ve for angles 0°–90° $\quad$ and $\quad$ 180°–270°

$\quad\quad$ − ve for angles 90°–180° $\quad$ and $\quad$ 270°–360°

4) When sketching the curve it is also worth remembering that at the '45° intervals' the tangent values are either +1 or −1.

Thus $\tan 45° = 1$, $\tan 135° = -1$, $\tan 225° = 1$, $\tan 315° = -1$

Inverse notation

Suppose that $\qquad\qquad \theta = 0.4771$

This means that θ is the angle whose sine is 0.4771

This may be written as $\qquad \theta = \text{inv} \sin 0.4771$

or $\qquad\qquad\qquad\qquad \theta = \text{arc} \sin 0.4771$

or $\qquad\qquad\qquad\qquad \theta = \sin^{-1} 0.4771$

These RHS expressions are known as inverse trigonometrical functions.

Finding angles over a specified range when given the sine, cosine or tangent

EXAMPLE 5.6

Find all the angles, between 0° and 360°, whose sines are 0.4771.

Our first move is to obtain a solution from our calculator using

$\boxed{\text{AC}}\ \boxed{0.4771}\ \boxed{\text{inv}}\ \boxed{\text{sin}}\ \underset{\text{DISPLAY}}{\boxed{28.496169}}$ giving 28.50° correct to 2 d.p.

Are there any other solutions? Well, this is where our knowledge of the trigonometrical waveforms will help.

Let us sketch the sine curve for angle values 0°–360°.

Now in a graph of $y = \sin \theta$ we need to consider when $y = 0.4771$ so we draw a horizontal line for an ordinate height of 0.4771 as shown in Fig. 5.28.

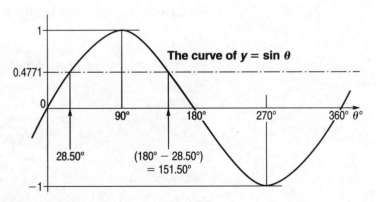

Fig. 5.28

At this stage you will have no difficulty in identifying the result from the calculator and also another angle whose sine is 0.4771. You can see from the symmetry of the curve how the other result is obtained. Thus 28.50° and 151.50° are the required angles.

EXAMPLE 5.7

Find all the angles between 0° and 360° whose cosines are −0.6354.

Again using our calculator we have

 giving 129.45° correct to 2 d.p.

This time we will sketch the curve of $y = \cos \theta$ for angle values 0°–360°, and then draw the horizontal line at ordinate height −0.6354 as shown in Fig. 5.29.

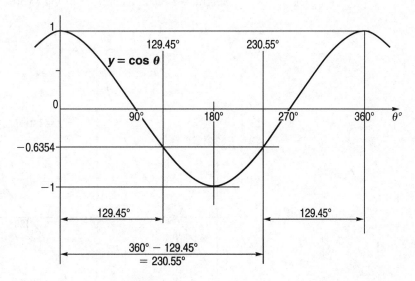

Fig. 5.29

Once again the diagram will explain how we use the symmetry of the waveform to find the value of the second angle. This is only one of many ideas which you could use.

Thus the required angles are   129.45°   and   230.55°.

EXAMPLE 5.8

Find all the angles between 0° and 360° whose tangents are −1.8972.

Using our calculator we have:

$\boxed{\text{AC}}$ $\boxed{1.8972}$ $\boxed{+/-}$ $\boxed{\text{inv}}$ $\boxed{\text{tan}}$ $\boxed{-62.206619}$ giving −62.21° correct to 2 d.p.
DISPLAY

This is interesting since this result from the calculator is outside the range of 0°–360°. We should not really be surprised since there are an infinite number of angles whose tangents are −1.8972. So that we can fit in this negative angle we will sketch the graph of $y = \tan \theta$ from −90° to 360°, and then draw the horizontal line at ordinate height −1.8972 as shown in Fig. 5.30.

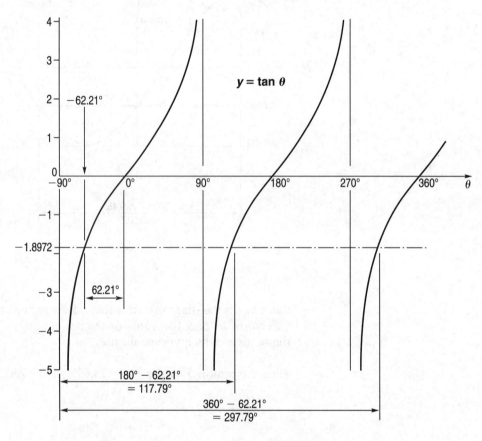

Fig. 5.30

This time it is slightly more difficult to see how the two required results are obtained, but symmetry is still important.

Hence the required angles are 117.79° **and** 297.79° **correct to 2 d.p.**

Exercise 5.2

1) Evaluate: $6\sin 23° - 2\cos 47° + 3\tan 17°$.

2) Evaluate: $5\sin 142° - 3\tan 148° + 3\cos 230°$.

3) Evaluate: $\sin A\ \cos B - \cos A\ \sin B$ given that $\sin A = \frac{3}{5}$ and $\tan B = \frac{4}{3}$. A and B are both acute angles. (*Hint*: sketch a right-angled triangle and show the given data on it.)

4) An angle A is in the 2nd quadrant. If $\sin A = \frac{3}{5}$ find, without actually finding angle A, the values of $\cos A$ and $\tan A$.

5) If $\sin\theta = 0.1432$ find all the values of θ from $0°$ to $360°$.

6) If $\cos\theta = -0.8927$ find all the values of θ from $0°$ to $360°$.

7) Find the angles in the first and second quadrants:

a) whose sine is 0.7137   b) whose cosine is -0.4813
c) whose tangent is 0.9476   d) whose tangent is -1.7642

8) Find the angles in the third or fourth quadrants:

a) whose sine is -0.7880 b) whose cosine is 0.5592
c) whose tangent is -2.9042

9) If $\sin A = \dfrac{a\sin B}{b}$ find the values of A between $0°$ and $360°$ when $a = 7.26$ mm, $b = 9.15$ mm   and $B = 18°29'$

Trigonometry and the circle

The circle is a figure that keeps cropping up in life generally, but especially in technology. Several references to the properties of the circle are made in the text of this book, but we think it is important enough to warrant study in its own right. This will also sharpen your awareness of the alternative measure of an angle, namely the radian.

Radian measure

We have seen that an angle is usually measured in degrees but there is another way of measuring an angle. In this, the unit is known as the radian (abbreviation rad).

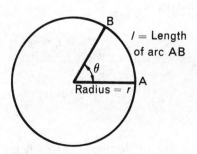

Fig. 5.31

Referring to Fig. 5.31 gives

$$\text{Angle in radians} = \frac{\text{Length of arc}}{\text{Radius of circle}}$$

$$\theta \text{ radians} = \frac{l}{r}$$

$$l = r\theta$$

Hence

$$\text{Length of arc} = r\theta$$

Relation between radians and degrees

If we make the arc AB equal to a semi-circle then

$$\text{Length of arc} = \pi r$$

and

$$\text{Angle in radians} = \frac{\pi r}{r} = \pi$$

Now the angle subtended by a semi-circle $= 180°$

Therefore

$$\pi \text{ radians} = 180°$$

or

$$1 \text{ radian} = \frac{180°}{\pi} = 57.3°$$

Thus to convert from degrees to radians

$$\theta° = \frac{\pi\theta}{180} \text{ radians}$$

Thus

$$30° = \frac{\pi(30)}{180} \text{ rad} = \frac{\pi}{6} \text{ rad}$$

$$90° = \frac{\pi}{2} \text{ rad} \qquad 180° = \pi \text{ rad}$$

$$45° = \frac{\pi}{4} \text{ rad} \qquad 270° = \frac{3\pi}{2} \text{ rad}$$

$$60° = \frac{\pi}{3} \text{ rad} \qquad 360° = 2\pi \text{ rad}$$

To convert from radians to degrees

$$\theta \text{ radians} = \left(\frac{180}{\pi} \times \theta\right)°$$

Degrees, minutes and seconds

There are	60 seconds in 1 minute, or	$60'' = 1'$
and	60 minutes in 1 degree, or	$60' = 1°$
Thus	60×60 seconds in 1 degree, or	$3600'' = 1°$

Modern calculating methods make the use of decimal degrees (e.g. $36.783°$) more likely than the use of minutes and seconds.

EXAMPLE 5.9

Convert $29°37'29''$ to radians stating the answer correct to 4 significant figures.

The first step is to convert the given angle into degrees and decimals of a degree.

$$29°37'29'' = 29 + \frac{37}{60} + \frac{29}{3600} = 29.625°$$

$$= \frac{\pi \times 29.625}{180} = 0.5171 \text{ radians}$$

Many scientific calculators will convert degrees, minutes and seconds into decimal degrees, and vice versa, using special keys—instructions for use of these keys will be given in the accompanying booklet.

EXAMPLE 5.10

Convert $0.089\,35$ radians into degrees, minutes and seconds.

$$0.089\,35 \text{ radians} = \frac{0.089\,35 \times 180}{\pi} = 5.1194°$$

For calculators without a decimal degree conversion facility the following sequence may be used—this is the reverse of the sequence used in the previous example.

Thus
$$\begin{aligned}
5.1194° &= (5 + 0.1194)° \\
&= 5° + (0.1194 \times 60)' \\
&= 5° + (7.164)' \\
&= 5° + (7 + 0.164)' \\
&= 5° + 7' + (0.164 \times 60)'' \\
&= 5° + 7' + (9.84)'' \\
&= 5° + 7' + 10'' \\
&= 5° \, 7' \, 10'' \quad \text{correct to the nearest second.}
\end{aligned}$$

Components of a circle

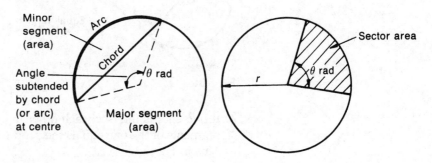

Fig. 5.32   Fig. 5.33

Useful properties of radii, chords and tangents

You will, no doubt, be familiar with most of the information given here, but it is included as most of these properties are used regularly in fine measurement calculations.

1) If two circles are tangential to each other then the straight line which passes through the centres of the circles also passes through the point of tangency.

   Thus the line AB joining the centres of the circles also passes through C, the point of tangency (Fig. 5.34) and AB is perpendicular to DE.

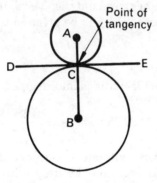

Fig. 5.34

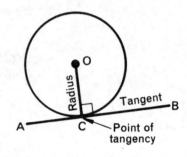

Fig 5.35

2) If a line is tangential to a circle then it is at right angles to a radius drawn to the point of tangency.

   Thus if AB is a tangent with C the point of tangency then the radius OC is at right angles to AB (Fig. 5.35).

3) If from a point outside a circle tangents are drawn to the circle, then their lengths are equal.

   Thus in Fig. 5.36 the lengths AC and BC are equal. It can also be proved that ∠ACB is bisected by CO, O being the centre of the circle.

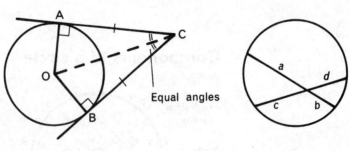

Fig. 5.36

Fig 5.37

4) If two chords intersect each other in a circle then the product of the segments of the one equals the product of the segments of the other. This in Fig. 5.37:

$$a \times b = c \times d$$

Area of a sector

The area of a circle $= \pi r^2$.

So, by proportion, (Fig. 5.33),

$$\text{Area of sector} = r^2 \times \frac{\theta}{2\pi}$$

$$= \tfrac{1}{2}r^2\theta$$

Summary

Length of arc $= r\theta$	θ in radians	or	$2\pi r\left(\dfrac{\theta^\circ}{360}\right)$
Area of a sector $= \dfrac{1}{2}r^2\theta$	θ in radians	or	$\pi r^2\left(\dfrac{\theta^\circ}{360}\right)$

EXAMPLE 5.11
Calculate a) the length of arc of a circle whose radius is 8 m and which subtends an angle of 56° at the centre, and b) the area of the sector so formed.

a) Length of arc $= 2\pi r \times \dfrac{\theta^\circ}{360} = 2 \times \pi \times 8 \times \dfrac{56}{360} = 7.82$ m

b) Area of sector $= \pi r^2 \times \dfrac{\theta^\circ}{360} = \pi \times 8^2 \times \dfrac{56}{360} = 31.28$ m^2

EXAMPLE 5.12
Find the angle of a sector of radius 35 mm and area 1020 mm^2

Now $\qquad$ Area of sector $= \tfrac{1}{2}r^2\theta$

and substituting the given values of

$$\text{Area} = 1020 \text{ mm}^2 \quad \text{and} \quad r = 35 \text{ mm}$$

we have $\qquad 1020 = \tfrac{1}{2}(35)^2\theta$

from which $\qquad \theta = \dfrac{1020 \times 2}{35^2} = 1.67$ rad

$$= \dfrac{180 \times 1.67}{\pi} = 95.7^\circ$$

EXAMPLE 5.13

Water flows in a 400 mm diameter pipe to a depth of 300 mm. Calculate the wetted perimeter of the pipe and the area of cross-section of the water.

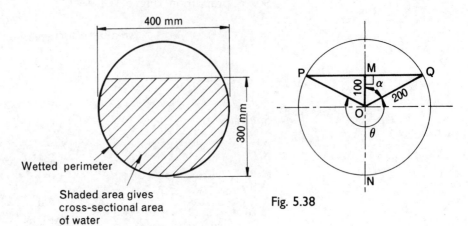

Wetted perimeter

Shaded area gives
cross-sectional area
of water

Fig. 5.38

From Fig. 5.38
the right-angled triangle MQO

$$\cos \alpha = \frac{OM}{OQ} = \frac{100}{200} = 0.5$$

$$\therefore \qquad \alpha = 60°$$

Also

$$\sin \alpha = \frac{MQ}{OQ}$$

$$\therefore \qquad MQ = OQ \sin \alpha = 200 \sin 60° = 173.2 \text{ mm}$$

Now

$$\theta + 2\alpha = 360°$$

$$\therefore \qquad \theta = 360° - 2(60°) = 240°$$

Thus

Wetted perimeter = Arc PNQ

$$= 2\pi r \left(\frac{\theta}{360} \right) = 2\pi(200) \left(\frac{240}{360} \right) = 838 \text{ mm}$$

Also

$$\left(\begin{array}{c} \text{Cross-sectional} \\ \text{area of water} \end{array} \right) = \left(\begin{array}{c} \text{Area of} \\ \text{sector PNQ} \end{array} \right) + \left(\begin{array}{c} \text{Area of} \\ \text{triangle POQ} \end{array} \right)$$

$$= \pi r^2 \left(\frac{\theta}{360} \right) + \tfrac{1}{2}(PQ)(MO)$$

$$= \pi(200)^2 \left(\frac{240}{360} \right) + \tfrac{1}{2}(2 \times 173.2)(100)$$

$$= 83\,780 + 17\,320$$

$$= 101\,000 \text{ mm}^2$$

Exercise 5.3

1) Convert the following angles to radians stating the answers correct to 4 significant figures:

   a) 35°   b) 83°28′   c) 19°17′32″   d) 43°39′49″

2) Convert the following angles to degrees, minutes and seconds correct to the nearest second:

   a) 0.1732 radians   b) 1.5632 radians   c) 0.0783 radians

3) If r is the radius and θ is the angle subtended by an arc, find the length of arc when:

   a) $r = 2\,\text{m}$, $\theta = 30°$ b) $r = 34\,\text{mm}$, $\theta = 38°40′$

4) If l is the length of an arc, r is the radius and θ the angle subtended by the arc, find θ when:

   a) $l = 9.4\,\text{m}$, $r = 4.5\,\text{m}$ b) $l = 14\,\text{mm}$, $r = 79\,\text{mm}$

5) An animal feed hopper incorporates an hinged shutter door which is in the shape of the sector of a circle. The length round the arc is 70 mm and the angle subtended at the centre is 45°. We need to know the length of one of the straight sides, which will be hinged, and in addition the area covered by the shutter.

6) The cross-section of a mechanical component is shown in Fig. 5.39. It is necessary to know the value of the shaded area.

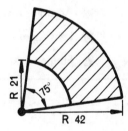

Fig. 5.39

7) We are trying to adapt an optical instrument to fit one of the items which our company manufactures. A chord 26 mm long has been ground on to the flat surface of a circular glass lens which has an effective diameter of 35 mm. You have been asked to calculate:

   a) the lengths of the arcs into which the effective circumference is divided.

   b) the area of the lens in the minor segment.

8) A flat is machined on a circular bar of 15 mm diameter, the maximum depth of cut being 2 mm. Find the area of the cross-section of the finished bar.

9) Water flows in a 300 mm diameter drain to a depth of 100 mm. Calculate the wetted perimeter of the drain and the area of cross-section of the water.

6 Solving practical engineering problems

The first part of this section will remind you of the theorem of Pythagoras and the basic trigonometrical ratios. This is followed by a selection of topics which are representative of those you may well meet in your career as an engineer. Remember these days you must be versatile and thus able to deal with a wide scope of problems, ranging from micro-chips to structural frameworks—but if you know your basic theory you will have no difficulty in solving the problem.

The theorem of Pythagoras

We must first remind ourselves that in a right-angled triangle the hypotenuse is the longest side and always lies opposite to the right-angle. Pythagoras' theorem states:

> In a right-angled triangle, the square on the hypotenuse is equal to the sum of the squares on the other two sides.

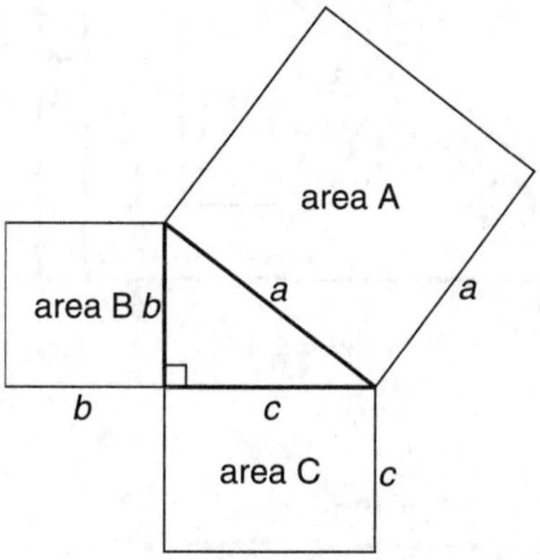

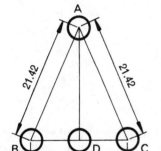

Fig. 6.1

A right-angled triangle is shown in Fig. 6.1 with the squares drawn on the sides. According to Pythagoras:

$$\text{area A} = \text{area B} + \text{area C}$$

or

$$a^2 = b^2 + c^2$$

For example, if $b = 3$ and $c = 4$

then

$$a^2 = 3^2 + 4^2$$
$$= 9 + 16 = 25$$

giving

$$a = 5$$

It is worth remembering that a triangle with sides 3, 4 and 5 is right angled. You may have to mark out a large right-angle on the ground, perhaps for the base of a garage—a protractor or set-square will not be accurate enough, but you can make the 3, 4, 5 triangle to as large a scale as you wish.

Triangles with sides 5, 12, 13 and 7, 24, 25 are also right-angled.

EXAMPLE 6.1

Four holes are bored in a plate as shown in Fig. 6.2. If D is mid-way between B and C, find the distance between A and D.

Now $\triangle$ABC is isosceles since it has two equal sides. The line AD bisects the base and is therefore perpendicular to BC.

Therefore $\triangle$ACD is right-angled and thus:

$$AC^2 = CD^2 + AD^2$$

or

$$AD^2 = AC^2 - CD^2 = 21.42^2 - 9.29^2 = 458.8 - 86.3$$
$$= 372.5$$

$$\therefore \quad AD = \sqrt{372.5} = 19.30 \text{ mm}$$

Fig. 6.2

Exercise 6.1

1) Two holes are bored in a plate to the dimensions shown in Fig. 6.3. To check the holes dimension m is required. What is this dimension?

2) Fig. 6.4 shows part of a drawing. If the holes are drilled correctly, what should be dimension x?

3) Fig. 6.5 shows a round bar of 30 mm diameter which has a flat milled on it. Find the width of the flat.

4) Fig. 6.6 shows a bar which has two opposite flats milled on it. Find the diatance d between the flats.

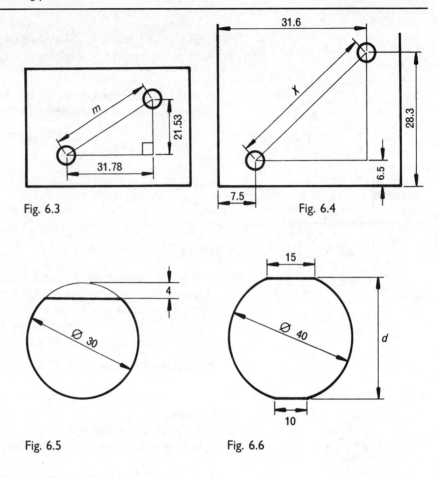

Fig. 6.3

Fig. 6.4

Fig. 6.5

Fig. 6.6

Basic trigonometrical ratios

For the definitions of these ratios we shall use a right-angled triangle labelled with standard notation, as shown in Fig. 6.7, Here angles are labelled A, B and C (in any order) and the sides are labelled a, b and c with each side opposite its corresponding angle.

The definitions of the three trigonometrical ratios are given below. Refer to Fig. 6.7.

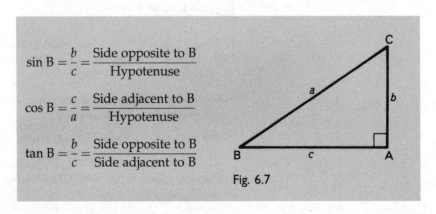

$$\sin B = \frac{b}{c} = \frac{\text{Side opposite to B}}{\text{Hypotenuse}}$$

$$\cos B = \frac{c}{a} = \frac{\text{Side adjacent to B}}{\text{Hypotenuse}}$$

$$\tan B = \frac{b}{c} = \frac{\text{Side opposite to B}}{\text{Side adjacent to B}}$$

Fig. 6.7

EXAMPLE 6.2

Find the sides marked x in Figs. 6.8, 6.9 and 6.10 correct to 3 s.f.

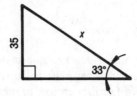

$$\sin 33° = \frac{35}{x}$$

$$x = \frac{35}{\sin 33°} = \frac{35}{0.545} = 64.3 \text{ mm}$$

Fig. 6.8

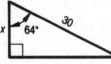

$$\frac{x}{30} = \cos 64°$$

$$x = 30 \times \cos 64°$$

$$= 30 \times 0.4384 = 13.2 \text{ mm}$$

Fig. 6.9

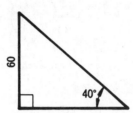

$$\tan 40° = \frac{60}{x}$$

$$x = \frac{60}{\tan 40°} = \frac{60}{0.839} = 71.5 \text{ mm}$$

Fig. 6.10

EXAMPLE 6.3

Find the angles marked θ in Figs 6.11 and 6.12 correct to 3 s.f.

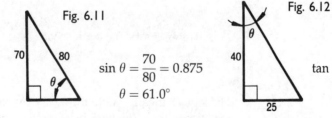

Fig. 6.11

$$\sin \theta = \frac{70}{80} = 0.875$$

$$\theta = 61.0°$$

Fig. 6.12

$$\tan \theta = \frac{25}{40} = 0.625$$

$$\theta = 32.0°$$

Reciprocal ratios

In addition to sin, cos and tan there are three other ratios that may be obtained from a right-angled triangle. These are:

cosecant	(called cosec for short)
secant	(called sec for short)
cotangent	(called cot for short)

The three ratios are defined as follows:

$$\operatorname{cosec} A = \frac{1}{\sin A} \qquad \sec A = \frac{1}{\cos A} \qquad \cot A = \frac{1}{\tan A}$$

The reciprocal of x is $\dfrac{1}{x}$ and it may therefore be seen why the terms cosec, sec and cot are called reciprocal ratios', since they are equal respectively to $\dfrac{1}{\sin}$, $\dfrac{1}{\cos}$ and $\dfrac{1}{\tan}$.

Formulae in technical reference books often include reciprocal ratios. It will then be necessary for you to re-write the formula before use in terms of the more familiar ratios, namely sin, cos, and tan.

For example, the formula used for checking the form of a metric thread:

$$M = D - \frac{5p}{6}\cot\theta + d(\operatorname{cosec}\theta + 1)$$

should be re-written as

$$M = D - \frac{5p}{6}\left(\frac{1}{\tan\theta}\right) + d\left(\frac{1}{\sin\theta} + 1\right)$$

Trigonometric identities

A statement of the type $\operatorname{cosec} A \equiv \dfrac{1}{\sin A}$ is called an *identity*.

The sign $\equiv$ means 'is identical to'. Any statement using this sign is true for all values of the variables, i.e. the angle A in the above identity. In practice, however, the $\equiv$ sign is often replaced by the $=$ (equals sign) and the identity would be given as $\operatorname{cosec} A = \dfrac{1}{\sin A}$.

Many trigonometrical identities may be verified by the use of a right-angled triangle.

EXAMPLE 6.4

To show that
$$\tan A = \frac{\sin A}{\cos A}$$

The sides and angles of the triangle may be labelled in any way providing that the 90° angle is *not* called A, (Fig. 6.13).

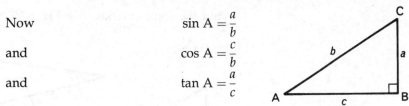

Now $\qquad \sin A = \dfrac{a}{b}$

and $\qquad \cos A = \dfrac{c}{b}$

and $\qquad \tan A = \dfrac{a}{c}$

Fig. 6.13

Hence from the given identity,

$$\text{RHS} = \frac{\sin A}{\cos A} = \frac{a/b}{c/b} = \frac{ab}{bc} = \frac{a}{c} = \tan A = \text{LHS}$$

EXAMPLE 6.5

To show that $\sin^2 A + \cos^2 A = 1$

In Fig. 6.13 $\qquad \sin A = \dfrac{a}{b} \qquad \therefore \quad \sin^2 A = \left(\dfrac{a}{b}\right)^2 = \dfrac{a^2}{b^2}$

$\qquad\qquad\qquad \cos A = \dfrac{c}{b} \qquad \therefore \quad \cos^2 A = \left(\dfrac{c}{b}\right)^2 = \dfrac{c^2}{b^2}$

$\therefore \qquad\qquad\qquad \text{LHS} = \sin^2 A + \cos^2 A = \dfrac{a^2}{b^2} + \dfrac{c^2}{b^2} = \dfrac{a^2 + c^2}{b^2}$

But by Pythagoras' theorem, $a^2 + c^2 = b^2$

$\therefore \qquad\qquad\qquad \text{LHS} = \dfrac{b^2}{b^2} = 1 = \text{RHS}$

Thus $\qquad\qquad\qquad \sin^2 A + \cos^2 A = 1$

Exercise 6.2

Find the lengths of the sides marked x in Fig. 6.14.

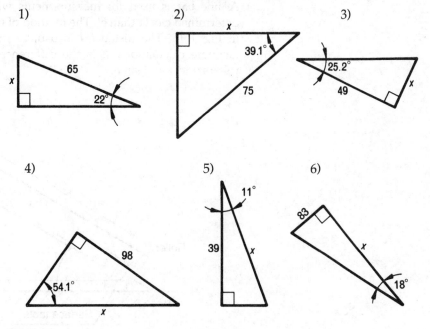

Fig. 6.14

Find the angles marked θ in Fig. 6.15.

7)   8)   9)

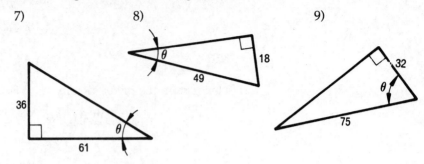

Fig. 6.15

10) The altitude of an isosceles triangle is 86 mm and each of the equal angles is 29°. Calculate the lengths of the equal sides.

Accurate measurements in manufacture

Here we have the application of basic trigonometric functions to problems faced by the fine measurement engineer or perhaps the production quality control inspector.

The sine bar

A sine bar is used for measurements which require the angle to be determined closer than 5′. The method of using the instrument is shown in Fig. 6.16. The distance l is usually made 100 mm or 200 mm to facilitate calculations. h is the difference in height between the two rollers and $h = l \sin \theta$.

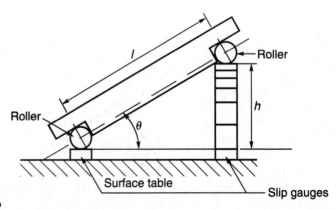

Fig. 6.16

EXAMPLE 6.6

The angle of 15°42′ is to be checked on the metal block shown in Fig. 6.17. Find the difference in height between the two rollers which support the ends of the 200 mm sine bar.

Now

$h = l \sin \theta$

$\quad = 200 \times \sin 15°42′$

$\quad = 200 \times 0.2706$

$\quad = 54.12 \text{ mm}$

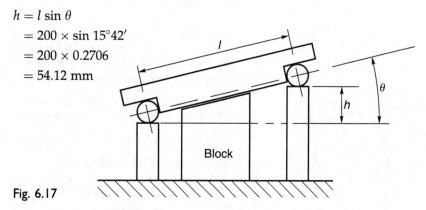

Fig. 6.17

The difference in height of the slip gauges must therefore be 54.12 mm if the angle is correct.

Exercise 6.3

1) Calculate the setting of a 100 mm sine bar to measure an angle of 27°15′.

2) Find the setting of a 200 mm sine bar to check a taper piece which has a taper of 1 in 10 on diameter. The piece is mounted in a similar way to that shown in Fig. 6.19.

3) Calculate the setting of a 200 mm sine bar to check a taper of 1 in 8 on diameter.

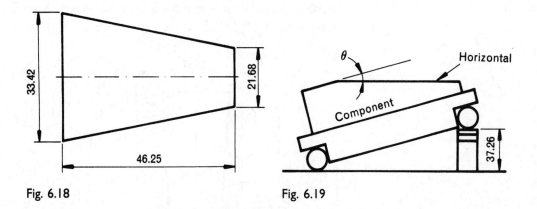

Fig. 6.18

Fig. 6.19

4) Find the setting of a 100 mm sine bar to check the taper piece shown in Fig. 6.18. The taper piece is mounted in a similar way to the component shown in Fig. 6.19.

5) Fig. 6.19 shows a component mounted on a 200 mm sine bar. Calculate the angle θ.

Reference rollers and balls

Sets of rollers can be obtained which are guaranteed to be within 0.002 mm for both diameter and roundness. By using rollers and balls many problems in measurement can be solved, some examples of which are given below.

EXAMPLE 6.7

A taper angle of 7° is to be checked by means of an adjustable gauge. The gauge is to be set by means of two rollers, one of 20 mm diameter and the other of 25 mm diameter. Find the centre distance *l* between the rollers.

In Fig. 6.20 A and B are the centres of the rollers. E and D are the points where the rollers touch the top blade.

$\angle AED = \angle BDE = 90°$ (angles between a radius and a tangent)

Now draw AC parallel to ED,

In $\triangle CAB$, $\angle CAB = 3°30'$ (half angle of taper)

$\angle ACB = 90°$,

$BC = BD - AE = 12.5 - 10 = 2.50$ mm

$AB = l$

$\therefore \dfrac{l}{2.50} = \text{cosec } 3°30'$

$l = 2.50 \times \text{cosec } 3°30' = 40.95$ mm

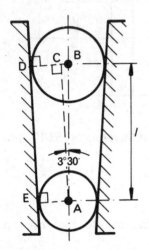

Fig. 6.20

EXAMPLE 6.8

A taper piece has a taper of 1 in 8 on the diameter. Two pairs of rollers 15.00 mm in diameter are used to check the taper as shown in Fig. 6.21. The measurement over the top rollers is 55.87 mm. Find:

a) the measurement over the bottom rollers if the taper is correct,
b) the bottom diameter of the job.

The first step is to find the angle of the taper. Using Fig. 6.22 we see that $\tan \alpha = \dfrac{0.5}{8}$ so $\alpha = 3°34'$.

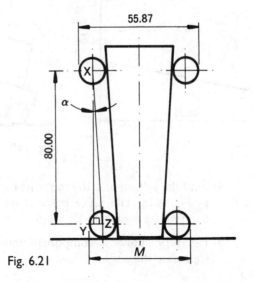

Fig. 6.21

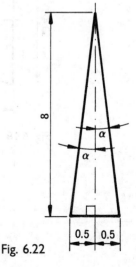

Fig. 6.22

a) Referring to Fig. 6.21, we have in $\triangle XYZ$ that:

$$\angle\alpha = 3°34', \quad XY = 80.00 \text{ mm}$$

but
$$\frac{YZ}{XY} = \tan\alpha$$

$\therefore \qquad YZ = 80 \times \tan 3°34' = 5.00$

Thus $\qquad M = 55.87 - 2 \times 5.00 = 45.87 \text{ mm}$

b) Referring to Fig. 6.23, we need to find AB.

$$\angle ABC = 90° - 3°34' = 86°26'$$

Since AB and BC are tangents, the line BZ bisects $\angle ABC$.

Hence $\qquad ABZ = 43°13'$

In $\triangle ABZ \qquad \dfrac{AB}{AZ} = \cot 43°13'$

$\therefore \qquad AB = AZ(\cot 43°13')$

$\qquad\qquad\quad = 7.5(\cot 43°13') = 7.982 \text{ mm}$

$\therefore$ Bottom diameter $= M - (2 \times AB) - (2 \times \text{radius of roller})$

$\qquad\qquad = 45.87 - 7.982 - 2 \times 7.50 = 14.91 \text{ mm}$

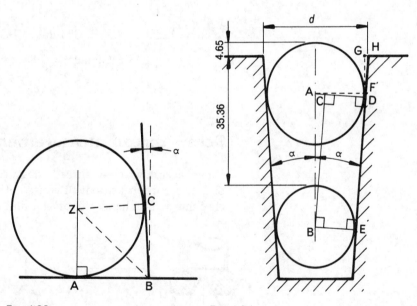

Fig. 6.23   Fig. 6.24

EXAMPLE 6.9

The tapered hole shown was inspected by using two balls 25.00 and 20.00 mm diameter, respectively. The measurements indicated in Fig. 6.24 were obtained. Find:

a) the included angle of taper 2α,

b) the top diameter d of the hole.

a) In the Fig. 6.24, A and B are the centres of the balls and E and D are points where the balls just touch the sides of the hole.

$$\angle ADE = \angle BED = 90° \quad \text{(angles between radius and tangent)}$$

Draw BC parallel to DE; then in $\triangle ABC$

$$AC = 12.50 - 10.00 = 2.50 \text{ mm}$$
$$AB = 35.36 + 4.65 - 12.50 + 10.00 = 37.51 \text{ mm}$$

Now $\quad \angle ACB = 90° \quad$ and $\quad \angle ABC = \alpha$

$$\sin \alpha = \frac{AC}{AB} = \frac{2.50}{37.51} \qquad \therefore \quad \alpha = 3°49'$$

$\therefore \qquad$ Included angle of taper $= 2\alpha = 2 \times 3°49' = 7°38'$.

b) To find d, draw AF horizontal and FG vertical.

Then $\qquad\qquad\qquad d = 2 \times (AF + GH)$

In $\triangle AFD$,

$$AD = 12.50 \text{ mm}, \quad \angle FAD = \alpha, \quad \angle ADF = 90°$$
$$\therefore \qquad AF = AD \sec \alpha = 12.50 \times \sec 3°49' = 12.53 \text{ mm}$$

In $\triangle GFH$,

$$GF = 12.50 - 4.65 = 7.85 \text{ mm}, \quad \angle FGH = 90°, \quad \angle GFH = \alpha$$
$$\therefore \qquad GH = GF \tan \alpha = 7.85 \times \tan 3°49' = 0.52 \text{ mm}$$

Thus $\qquad d = 2(AF + GH) = 2 \times (12.53 + 0.52) = 26.10 \text{ mm}$

Screw thread measurement

When accurate screw thread measurement is required the method of 2 or 3 wire measurement is used. The 3-wire method is used when checking with a hand micrometer and the 2-wire method is used with

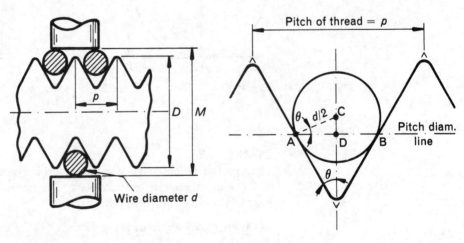

Fig. 6.25   Fig. 6.26

bench micrometers. The methods are fundamentally the same and allow the pitch or effective diameter to be measured (Fig. 6.25).

For the most accurate results each type and size of thread requires wires of a certain size. The best wire size is one which just touches the thread flanks at the pitch diameter as shown in Fig. 6.26.

At the pitch line diameter the distance between the flanks of the thread is equal to half the pitch.

Hence in Fig. 6.26

$$AB = \frac{p}{2} \quad \text{and} \quad AD = \frac{p}{4}$$

In $\triangle ADC$, $\dfrac{AC}{AD} = \sec\theta$

$\therefore$ $AC = AD \times \sec\theta$

$\therefore$ $\dfrac{d}{2} = \dfrac{p}{4}\sec\theta \qquad \therefore \quad d = \dfrac{p}{2}\sec\theta$

For a metric thread, $\theta = 30°$ and hence the best wire size is

$$d = \frac{p}{2}\sec 30° = 0.5774p$$

Formula for checking the form of a metric thread

Although the best wire size should be used, in practice the wire used will vary a little from this best size. In order to determine the distance over the wires for a specific thread (M in Fig. 6.25) the following formula is used

$$M = D - \frac{5p}{6}\cot\theta + d(\operatorname{cosec}\theta + 1)$$

For a metric thread, the formula becomes

$$M = D - \frac{5p}{6}\cot 30° + d(\operatorname{cosec} 30° + 1)$$
$$= D - 1.4434p + d(2 + 1)$$
$$= D - 1.4434p + 3d$$

EXAMPLE 6.10
A metric thread having a major diameter of 20 mm and a pitch of 2.5 mm is to be checked using the best wire size for this particular thread. Find:

a) the best wire size,
b) the measurement over the wires if the thread is correct.

a) The best wire size is

$$d = 0.5774p = 0.5774 \times 2.5 = 1.4435 \text{ mm}$$

b) The measurement over the wires is

$$M = D - 1.4434p + 3d$$
$$= 20 - 1.4434 \times 2.5 + 3 \times 1.4435 = 20.722 \text{ mm}$$

Exercise 6.4

1) A steel ball 40 mm in diameter is used to check the taper hole, a section of which is shown in Fig. 6.27. If the taper is correct, what is the dimension *x*?

2) A taper plug gauge is being checked by means of reference rollers and slip gauges. The set-up is as shown in Fig. 6.28. Find the included angle of the taper of the gauge and also the top and bottom diameters.

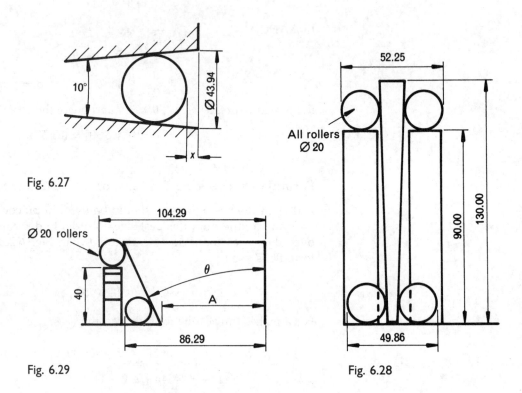

Fig. 6.27

Fig. 6.29

Fig. 6.28

3) Fig. 6.29 shows a dovetail being checked by rollers and slip gauges. Find the angle θ and the dimension *A*.

4) Calculate the dimension *M* which is needed for checking the groove, a cross-section of which is shown in Fig. 6.30.

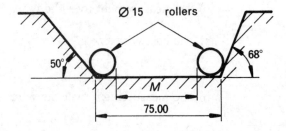

Fig. 6.30

5) A tapered hole has a maximum diameter of 32.00 mm and an included angle of 16°. A ball having a diameter of 20 mm is placed in the hole. Calculate the distance between the top of the hole and the top of the ball.

6) Fig. 6.31 shows the dimensions obtained in checking a tapered hole. Find the included angle of taper of the hole and the top diameter d.

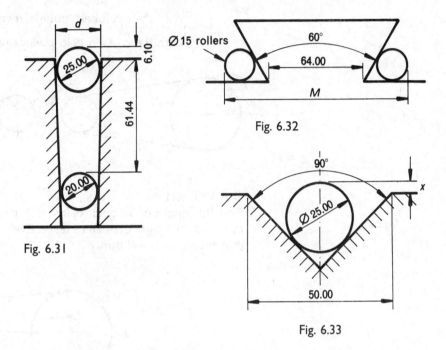

Fig. 6.31

Fig. 6.32

Fig. 6.33

7) Find the checking dimension M for the symmetrical dovetail slide shown in Fig. 6.32.

8) Fig. 6.33 shows a Vee block being checked by means of a reference roller. If the block is correct what is the dimension x?

9) A metric thread having a major diameter of 52 mm and a pitch of 5 mm is to be checked by the 3-wire method. Determine the best size of wire and, using the best wire size, determine the measurement over the wires if the thread is correct.

10) A metric thread having a major diameter of 30 mm and a pitch of 2 mm is to be checked using wires whose diameters are 1.14 mm. Calculate the measurement over the wires that will be obtained if the thread is correct.

Lengths of belts

Although gears are usually preferred for power transmission, belts (generally Vee belts) are used in applications where a more flexible drive is needed since they absorb and smooth out shock loading.

There are two distinct cases, open belts and crossed belts, as shown in Fig. 6.34.

With the open belt, pulleys revolve in the *same* direction.

With the crossed belt, pulleys revolve in *opposite* directions.

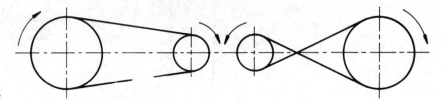

Fig. 6.34

EXAMPLE 6.11

Find the length of an open belt which passes over two pulleys of 200 mm and 300 mm diameter, respectively. The distance between the pulley centres is 900 mm.

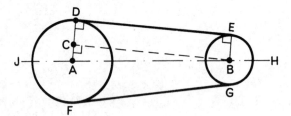

Fig. 6.35

Referring to Fig. 6.35, the total length of the belt is made up of the two straight lengths DE and FG and the arcs DJF and EHG.

$$\angle ADE = \angle BED = 90° \quad \text{(angle between radius and tangent)}$$

Draw CB parallel to DE. Then in $\triangle ABC$,

$$AC = AD - EB = 150 - 100 = 50 \text{ mm}$$
$$AB = 900 \text{ mm (given)}, \quad \angle ACB = 90°$$

$$\therefore \quad \cos CAB = \frac{AC}{AB} = \frac{50}{900} = 0.0556 \quad \therefore \quad \angle CAB = 86°49'$$

Now $\qquad BC = AB \sin CAB = 900 \times \sin 86°49' = 898.6 \text{ mm}$

Also $\qquad \angle EBH = \angle CAB = 86°49'$

Hence the arc EHG subtends an angle of $2 \times 86°49' = 173°38'$ at the centre.

$$\text{Length of arc EHG} = 2\pi \times 100 \times \frac{173°38'}{360°} = 303.0 \text{ mm}$$

Now $\quad\angle DAJ = 180° - \angle CAB = 180° - 86°49' = 93°11'$

The arc DJF therefore subtends an angle of $2 \times 93°11' = 186°22'$ at the centre.

$$\text{Length of arc DJF} = \frac{2\pi \times 150 \times 186°22'}{360°} = 488.0 \text{ mm}$$

$\therefore\quad$ Total length of belt $= 2 \times BC + \text{Arc EHG} + \text{Arc DJF}$
$$= 2 \times 898.6 + 303.0 + 488.0 = 2588 \text{ mm}$$

EXAMPLE 6.12

Two pulleys 200 mm and 300 mm in diameter respectively are placed 1200 mm apart. They are connected by a closed belt. Find the length of the belt required.

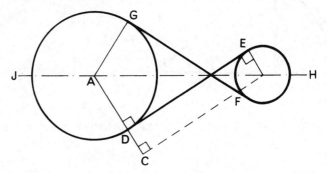

Fig. 6.36

The length of the belt is made up of two straight lengths ED and FG, arc GJD and arc EHF, as shown in Fig. 6.36.

$$\angle ADE = \angle DEB = 90° \quad \text{(angle between radius and tangent)}$$

Draw CB parallel to DE; then in $\triangle ABC$

$$AC = 100 + 150 = 250 \text{ mm}, \quad AB = 1200 \text{ mm}, \quad \angle ACB = 90°,$$

$$\therefore \quad \cos CAB = \frac{AC}{AB} = \frac{250}{1200} = 0.2083 \qquad \therefore \quad \angle CAB = 77°59'$$

Thus $\quad BC = AB \times \sin CAB = 1200 \times \sin 77°59' = 1174 \text{ mm}$

Also $\qquad\qquad\qquad \angle EBA = \angle CAB = 77°59'$
$\therefore \qquad\qquad\qquad \angle EBH = 180° - 77°59' = 102°1'$

The arc EHF therefore subtends an angle $2 \times 102°1'$ at the centre.

$$\therefore \quad \text{Length of arc EHF} = 2\pi \times 100 \times \frac{204°2'}{360°} = 356 \text{ mm}$$

Similarly

$$\text{Length of arc GJD} = 2\pi \times 150 \times \frac{204°2'}{360°} = 534 \text{ mm}$$

$\therefore \quad$ Total length of belt $= 2 \times 1174 + 356 + 534 = 3238 \text{ mm}$

Frameworks

Frameworks hardly need an introduction as we can all think of situations where they are used in engineering, although most of these would be for building and construction work. Typical mechanical engineering uses are in crane frameworks and motor vehicle construction.

EXAMPLE 6.13

Fig. 6.37 shows a framework. Calculate the lengths of the members BC, BD and AC.

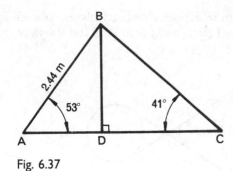

Fig. 6.37

In $\triangle$ABD,

$$\frac{BD}{AB} = \sin 53°$$

$\therefore$ $BD = AB \sin 53° = 2.44 \sin 53° = 1.95$

In $\triangle$BCD,

$$\frac{BC}{BD} = \operatorname{cosec} 41°$$

$\therefore$ $BC = BD \operatorname{cosec} 41° = 1.95 \operatorname{cosec} 41° = 2.97$

To find the length of AC we must first find the lengths of AD and DC.

In $\triangle$ABD, $\dfrac{AD}{AB} = \cos 53°$

$\therefore$ $AD = AB \cos 53° = 2.44 \cos 53° = 1.47$

In $\triangle$BCD, $\dfrac{DC}{BC} = \cos 41°$

$\therefore$ $DC = BC \cos 41° = 2.97 \cos 41° = 2.24$

Thus $AC = AD + DC = 1.47 + 2.24 = 3.71$

Hence BD is 1.95 m, BC is 2.97 m and AC is 3.71 m.

Exercise 6.5

1) A belt passes over a pulley 1200 mm in diameter. The angle of contact between the pulley and the belt is 230°. Find the length of belt in contact with the pulley.

2) An open belt passes over two pulleys 900 mm and 600 mm in diameter, respectively. If the centres of the pulleys are 1500 mm apart, find the length of the belt required.

3) Two pulleys of diameters 1400 mm and 900 mm, respectively, with centres 4.5 m apart, are connected by an open belt. Find its length.

4) An open belt connects two pulleys of diameters 120 mm and 300 mm with centres 300 mm apart. Calculate the length of the belt.

5) A crossed belt passes over two pulleys each of 450 mm diameter. If their centres are 600 mm apart, calculate the length of the belt.

6) **If, in Question 2, a crossed belt is used, what will be its length?**

7) **A crossed belt passes over two pulleys 900 mm and 1500 mm in diameter, respectively, which have their centres 6 m apart. Find its length.**

8) A crossed belt passes over two pulleys, one of 280 mm diameter and the other of 380 mm diameter. The angle between the straight parts of the belt is 90°. Find the length of the belt.

9) A framework, to be used horizontally for supporting a machine tool lifting device, is shown in Fig. 6.38 as triangle ABC. Calculate the length AC.

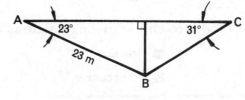

Fig. 6.38

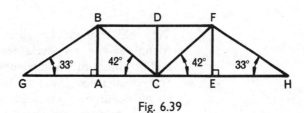

Fig. 6.39

10) The framework shown in Fig. 6.39 has been designed for supporting an overhead gantry rail type transfer system for use in the construction of heavy earth moving equipment. In order to assess the total length of material required, and also the individual lengths, you have been asked to calculate the lengths of all the members. You have been told that AB = CD = EF = 2.4 m.

'Any old triangle' is called, in mathematics, a scalene triangle—a triangle having no special features: no right angle, all angles different and no sides of the same length. This chapter covers the use of the sine and cosine rules. These are the two rules necessary for solving scalene triangles, unless you wish to break them down into right-angled triangle components. Direct practical applications include an example of a simple crane framework and a electrical phasor diagram.

When we have found the three missing elements in the solution of a triangle problem we are said to have 'solved the triangle'.

The sine rule

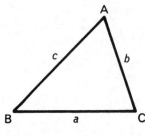

Fig. 7.1

The sine rule may be used when either of the following is known:

1) one side and any two angles; or

2) two sides and an angle opposite to one of these sides. (In this case two solutions may be found giving rise to what is called the 'ambiguous case', see Example 7.2.)

Using the notation of Fig. 7.1 the sine rule states:

$$\frac{a}{\sin A} = \frac{b}{\sin B} = \frac{c}{\sin C}$$

EXAMPLE 7.1
Solve the triangle ABC given that A = 42°, C = 72° and b = 61.8 mm.

The triangle should be drawn for reference as shown in Fig. 7.2 but there is no need to draw it to scale.

Since $\angle A + \angle B + \angle C = 180°$

then $\qquad\qquad \angle B = 180° - 42° - 72° = 66°$

The sine rule states:

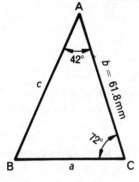

Fig. 7.2

$$\frac{a}{\sin A} = \frac{b}{\sin B} \qquad\qquad \frac{c}{\sin C} = \frac{b}{\sin B}$$

$$\therefore \quad a = \frac{b \sin A}{\sin B} \qquad\qquad \therefore \quad c = \frac{b \sin C}{\sin B}$$

$$= \frac{61.8 \times \sin 42°}{\sin 66°} \qquad\qquad = \frac{61.8 \times \sin 72°}{\sin 66°}$$

$$= 45.3 \,\text{mm} \qquad\qquad\qquad = 64.3 \,\text{mm}$$

The complete solution is:

$$\angle B = 66°, a = 45.3\,\text{mm}, c = 64.3\,\text{mm}$$

A rouch check on sine rule calculations may be made by remembering that in any triangle the longest side lies opposite the largest angle and the shortest side lies opposite the smallest angle.

Thus in the previous example:

Smallest angle = 42° = A; Shortest side = a = 45.3 mm
Largest angle = 72° = C;   Longest side = c = 64.3 mm

The ambiguous case

Look at these two triangles:

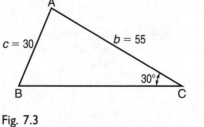 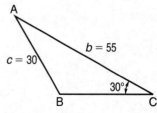

Fig. 7.3   Fig. 7.4

and now consider the next example.

EXAMPLE 7.2

Solve $\triangle ABC$ if $b = 55\,\text{mm}$, $c = 30\,\text{mm}$ and $\angle C = 30°$.

Since we are given two sides and an angle, we look to the sine rule.

So using
$$\frac{b}{\sin B} = \frac{c}{\sin C}$$

from which
$$\sin B = \frac{b \sin C}{c}$$

$$= \frac{55 \sin 30°}{30} = 0.9167$$

giving
$$\angle B = 66.4°$$

Now, no doubt, you will have noticed that the given data would fit either of the triangles, Fig. 7.3 or Fig. 7.4. This is why it is called the ambiguous case.

But how is it that we have worked through using trigonometry and found only one value for $\angle B$ of 66.4° which is clearly suitable for Fig. 7.3?

Where does the other value of $\angle B$ come from? In fact, there are two angles between 0° and 180° which have the same sine value. These are supplementary angles and therefore add up to 180°.

Hence the other value of
$$\angle B = 180° - 66.4°$$

$$= 113.6° \quad \text{which fits Fig. 7.4.}$$

Although your calculator will only produce one value of angle for a specified sine value, it is possible to verify that $\sin 113.6° = 0.9167$ (see Chapter 5 on inverse trigonometric ratios).

We will now find the full solutions for both triangles.

When	When
$\angle B = 66.4°$	$\angle B = 113.6°$
$\angle A = 180° - 66.4° - 30°$	$\angle A = 180° - 113.6° - 30°$
$= 83.6°$	$= 36.4°$
Now	Now
$\dfrac{a}{\sin A} = \dfrac{c}{\sin C}$	$\dfrac{a}{\sin A} = \dfrac{c}{\sin C}$
$\therefore \quad a = \dfrac{c \sin A}{\sin C}$	$\therefore \quad a = \dfrac{c \sin A}{\sin C}$
$= \dfrac{30 \sin 83.6°}{\sin 30°}$	$= \dfrac{30 \sin 36.4°}{\sin 30°}$
$= 59.6\,\text{mm}$	$= 35.6\,\text{mm}$

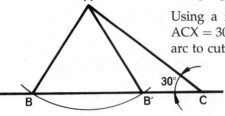

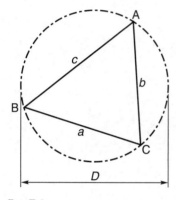

Fig. 7.5

Fig. 7.6

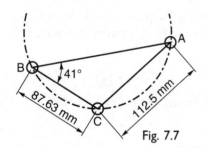

Fig. 7.7

The ambiguous case may be seen clearly by constructing the given triangle geometrically as follows (Fig. 7.5).

Using a full size scale draw AC = 55 mm and draw CX such that ACX = 30°. Now with centre A and radius 30 mm describe a circular arc to cut CX at B and B'.

Then ABC represents the triangle shown in Fig. 7.3 and AB'C represents the triangle shown in Fig. 7.4.

As long as you are aware of the two possible solutions, where this occurs in an engineering situation, there is almost always an indication as to which shape of triangle is required.

Use of the sine rule to find the diameter (*D*) of the circumscribing circle of a triangle

Using the notation of Fig. 7.6

$$\frac{a}{\sin A} = \frac{b}{\sin B} = \frac{c}{\sin C} = D$$

The rule is useful when we wish to find the pitch circle diameter of a ring of holes.

EXAMPLE 7.3

In Fig. 7.7 three holes are positioned by the angle and dimensions shown. Find the pitch circle diameter.

We are given $\angle B = 41°$ and $b = 112.5\,\text{mm}$

$$\therefore \qquad D = \frac{b}{\sin B} = \frac{112.5}{\sin 41°}$$

$$= 171.5\,\text{mm}$$

The cosine rule

The cosine rule is used in all cases where the sine rule cannot be used. These are when either of the following is known:

(1) two sides and the angle between them:
(2) three sides.

Whenever possible the sine rule is used because it results in a calculation which is easier to perform. In solving a triangle it is sometimes necessary to start with the cosine rule and then, having found one of the unknown elements, to finish solving the triangle using the sine rule.

The cosine rules states:	
either	$a^2 = b^2 + c^2 - 2bc \cos A$
or	$b^2 = a^2 + c^2 - 2ac \cos B$
or	$c^2 = a^2 + b^2 - 2ab \cos C$

EXAMPLE 7.4
Solve the triangle ABC if $a = 70\,\text{mm}$, $b = 40\,\text{mm}$ and $\angle C = 64°$.
Referring to Fig. 7.8, to find the side c we use

$$c^2 = a^2 + b^2 - 2ab \cos C$$

$$= 70^2 + 40^2 - 2 \times 70 \times 40 \times \cos 64°$$

$$\therefore \qquad c = \sqrt{4044} = 63.6 \text{ mm}$$

We now use the sine rule to find $\angle A$:

$$\frac{a}{\sin A} = \frac{c}{\sin C}$$

$$\sin A = \frac{a \sin C}{c} = \frac{70 \times \sin 64°}{63.6}$$

Thus $\qquad A = 81.6°$

and $\qquad B = 180° - 81.6° - 64° = 34.4°$

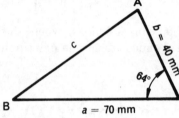

Fig. 7.8

EXAMPLE 7.5
The mast AB of a jib crane (Fig. 7.9) is 3 m long and the tie BC is 2.4 m long. If the jib AC is 4.8 m long, find the angle B between the mast and the tie.

Here we shall use $\qquad\qquad b^2 = a^2 + c^2 - 2ac \cos B$

Since we require B, we shall make B the subject of the equation.

Adding $2ac \cos B$ to both sides:

$$2ac \cos B + b^2 = a^2 + c^2 - \cancel{2ac \cos B} + \cancel{2ac \cos B}$$

and subtracting b^2 from both sides:

$$2ac \cos B + \cancel{b^2} - \cancel{b^2} = a^2 + c^2 - b^2$$

and finally dividing both sides by $2ac$:

$$\frac{\cancel{2ac} \cos B}{\cancel{2ac}} = \frac{a^2 + c^2 - b^2}{2ac}$$

giving $\qquad\qquad \cos B = \dfrac{a^2 + c^2 - b^2}{2ac}$

Hence $\qquad\qquad \cos B = \dfrac{2.4^2 + 3^2 - 4.8^2}{2 \times 2.4 \times 3}$

$$= -0.5750$$

or $\qquad\qquad B = \text{inv } \cos(-0.5750)$

giving $\qquad\qquad B = 125°$

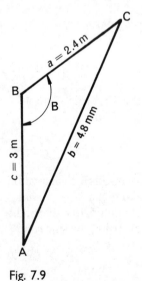

Fig. 7.9

EXAMPLE 7.6

The instantaneous values, i_1 and i_2, of two alternating currents are represented by the two sides of a triangle shown in Fig. 7.10. The resultant current i_R is represented by the third side. Calculate the magnitude of i_R and the angle ϕ between the current i_1 and i_R.

In $\triangle ABC$, Fig. 7.10, we have

$$b = 10, \ a = 15 \quad \text{and} \quad \angle C = 120°.$$

Using the cosine rule gives

$$c^2 = a^2 + b^2 - 2ac \cos C$$
$$= 15^2 + 10^2 - 2 \times 15 \times 10 \times \cos 120°$$
$$= 225 + 100 + 150$$
$$\therefore \qquad c = \sqrt{475} = 21.79 = i_R$$

To find $\angle A$ we use the sine rule,

$$\frac{a}{\sin A} = \frac{c}{\sin C}$$
$$\therefore \quad \sin A = \frac{a \sin C}{c} = \frac{15 \times \sin 120°}{21.79} = \frac{15 \times \sin 60°}{21.79} = 0.5962$$
$$\therefore \qquad A = 36.6° = \phi$$

Hence the magnitude of i_R is 21.8 and the angle ϕ is 36.6°.

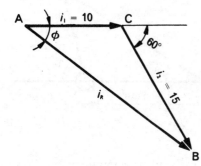

Fig. 7.10

Exercise 7.1

1) Solve the following triangles using the sine rule:
   a) $A = 75°$
      $B = 34°$
      $a = 102 \, \text{mm}$
   b) $C = 61°$
      $B = 71°$
      $b = 91 \, \text{mm}$

2) Solve the following triangles ABC using the cosine rule:
   a) $a = 9 \, \text{m}$
      $b = 11 \, \text{m}$
      $C = 60°$
   b) $b = 10 \, \text{m}$
      $c = 14 \, \text{m}$
      $A = 56°$

3) Three holes lie on a pitch circle and their chordal distances are 41.82 mm, 61.37 mm and 58.29 mm. Find their pitch circle diameter.

4) In Fig. 7.11, find the angle BCA given that BC is parallel to AD.

5) Calculate the angle θ in Fig. 7.12. There are 12 castellations and they are equally spaced.

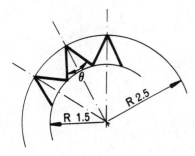

Fig. 7.11

Fig. 7.12

175

6) Find the smallest angle in a triangle whose sides are 20, 25 and 30 m long.

7) In Fig. 7.13 find:

   a) the distance AB

   b) the angle ACB

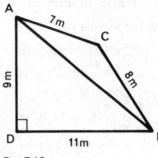

Fig. 7.13

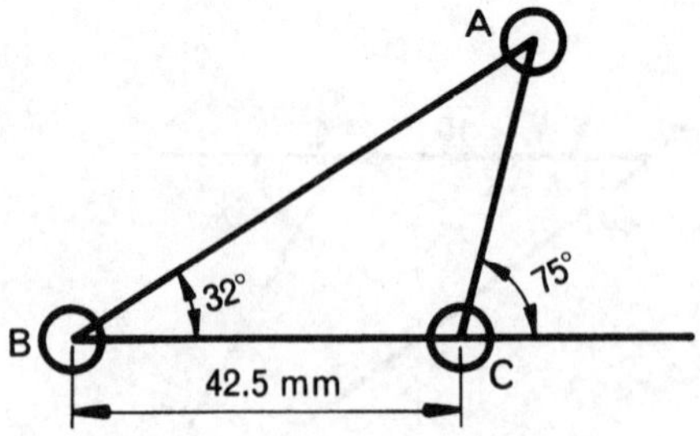

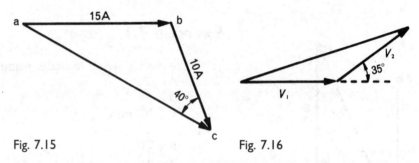

Fig. 7.14

8) Three holes are spaced in a plate detail as shown in Fig. 7.14. Calculate the centre distances from A to B and from A to C.

9) In Fig. 7.15, *ab* and *bc* are phasors representing the alternating currents in two branches of a circuit. The line *ac* represents the resultant current. Find by calculation this resultant current.

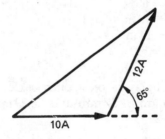

Fig. 7.15

Fig. 7.16

10) Two phasors are shown in Fig. 7.16. If $V_1 = 8$ and $V_2 = 6$ calculate the value of their resultant and the angle it makes with V_1.

11) Calculate the resultant of the two phasors shown in Figure 7.17.

Fig. 7.17

8 Area of a triangle

The area of a triangle is another important feature needed if we are to know all the facts relating to our basic triangular shape. Areas are needed when calculating the surface of fabricated components—they may be required in order that the mass of a part may be estimated. The costs of plating and also painting are very often based on the surface areas of items.

Three formulae are commonly used for finding the areas of triangles. These are used if we know any one of the following cases:

1) the base and the altitude (i.e. the 'height' perpendicular to the base)
2) any two sides and the included angle
3) the three sides.

Area given the base and the altitude

In Fig. 8.1

$$\text{Area of triangle} = \frac{1}{2} \times \text{Base} \times \text{Altitude}$$

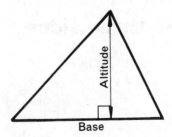

Fig. 8.1

177

EXAMPLE 8.1

Find the areas of the triangular shaped templates shown in Fig. 8.2. One purpose of this example is for you to appreciate that the 'base' side need not be horizontal. In each case the 'base' is chosen as the side of known length, and the altitude is measured at right angles to this side.

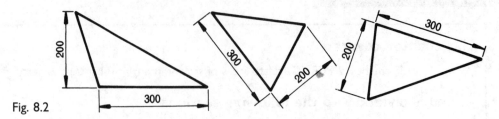

Fig. 8.2

$$\text{Template area} = \frac{1}{2} \times \text{Base} \times \text{Altitude} = \frac{1}{2} \times 300 \times 200 = 30\,000 \text{ mm}^2$$

Area given any two sides and the included angle

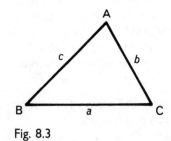

Fig. 8.3

In Fig. 8.3,

$$\text{Area of triangle} = \frac{1}{2}bc \sin A$$

or

$$\text{Area of triangle} = \frac{1}{2}ac \sin B$$

or

$$\text{Area of triangle} = \frac{1}{2}ab \sin C$$

EXAMPLE 8.2

The profile shown in Fig. 8.4 is that of a cutting blade for use in a carpet making machine. These blades will have to be electroplated when produced in quantity and the cost involved is largely based on the total surface area, neglecting the thickness of the blade.

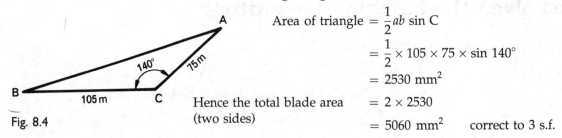

Fig. 8.4

$$\text{Area of triangle} = \frac{1}{2}ab \sin C$$

$$= \frac{1}{2} \times 105 \times 75 \times \sin 140°$$

$$= 2530 \text{ mm}^2$$

Hence the total blade area (two sides)

$$= 2 \times 2530$$

$$= 5060 \text{ mm}^2 \qquad \text{correct to 3 s.f.}$$

Area given the three sides

In Fig. 8.3

$$\text{Area of triangle} = \sqrt{s(s-a)(s-b)(s-c)}$$

where

$$s = \frac{a+b+c}{2}$$

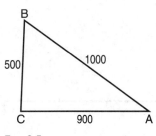

Fig. 8.5

EXAMPLE 8.3

The cover for a gearbox on a machine tool is to be formed from the sheet metal blank shown in Fig. 8.5. The production department need to know the mass of the blank if it is cut from material having a mass of 8 kg per square metre of surface area.

Now
$$s = \frac{a+b+c}{2}$$

Since we need the area in square metres, and also to keep the figures reasonably simple, we will work in metre units.

Thus
$$s = \frac{0.5+0.9+1.0}{2}$$
$$= 1.2$$

Now   Area of $\triangle ABC = \sqrt{(s-a)(s-b)(s-c)}$
$$= \sqrt{1.2(1.2-0.5)(1.2-0.9)(1.2-1.0)}$$
$$= \sqrt{1.2 \times 0.7 \times 0.3 \times 0.2}$$
$$= 0.2245 \text{ mm}^2$$

Thus   Mass of blank $= 0.2245 \times 8$
$$= 1.80 \text{ kg} \qquad \text{correct to 3 s.f.}$$

Exercise 8.1

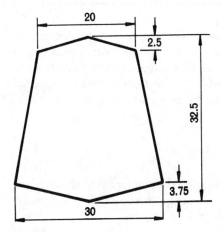

Fig. 8.6

1) We are about to go into production of portable notices to alert motorists of roadworks. The shape will essentially be that of an isosceles triangle with two equal angles of 50° and a base length of 450 mm. In order to estimate the cost of painting we need to know the total (back and front) surface area in square metre units.

2) A plate in the shape of an equilateral triangle has a mass of 12.25 kg. If the material has a mass of 3.7 kg/m², find the dimensions of the plate in millimetres.

3) Obtain the area of a triangular plastic sheet whose sides are 39.3 m and 41.5 m if the angle between them is 41°30′.

4) Find the area of the template shown in Fig. 8.6.

5) A small aluminium wedged shaped component (Fig. 8.7) is used in a patent levelling device for a precision instrument. Since the overall mass is of paramount importance we need to know how much extra mass is added to the instrument if there are four of these wedges per unit. Take the density of aluminium is 2700 kg/m³.

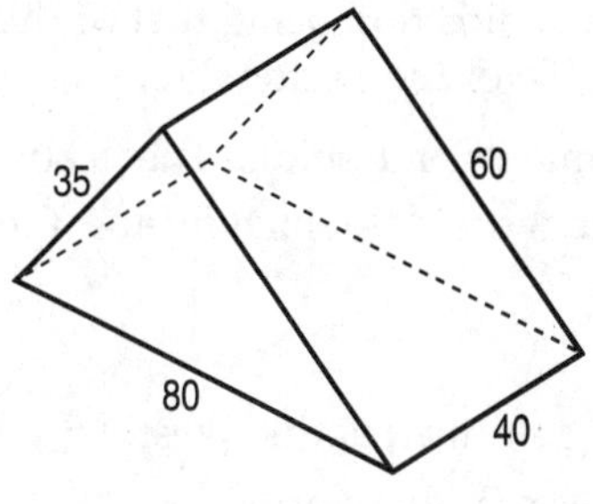

6) We have been informed that the wedges shown in Fig. 8.7 are 30% over the maximum permitted mass. It has therefore been decided that the top apex portion will be removed. This is to be achieved by a cut parallel to the 80 mm by 40 mm base. What will be the depth of the final component?

Fig. 8.7

Vectors

The work we cover on vectors only serves as a basic introduction, as vector analysis is an extensive subject in its own right. An application of vectors is basic navigation using 'velocity triangles'. Here velocities can be represented as vectors, in a similar way to forces. Phasors—special types of vectors—cater for electrical currents and voltages.

Scalars and vectors

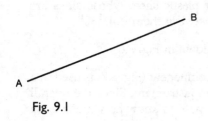

Fig. 9.1

A scalar quantity is one that is fully defined by magnitude alone. Some examples of scalar quantities are time (e.g. 30 seconds), temperature (e.g. 8 degrees Celsius) and mass (e.g. 7 kilograms).

A vector quantity needs magnitude, direction and sense to describe it fully. A vector may be represented by a straight line, its length representing the magnitude of the vector, its direction being that of the vector and suitable notation giving the sense of the vector (Fig. 9.1).

The usual way of naming a vector is to name its end points. The vector in Fig. 9.1 starts at A and ends at B and we write $\overrightarrow{AB}$ which means 'the vector from A to B'—this gives the sense.

Some examples of vector quantities are:

1) A displacement in a given direction, e.g. 15 metres due west.

2) A velocity in a given direction, e.g. 40 km/h due north.

3) A force of 20 kN acting vertically downwards.

Graphical representation

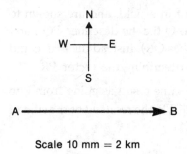

Scale 10 mm = 2 km

Fig. 9.2

A vector may be represented by a straight line drawn to scale.

EXAMPLE 9.1
A man walked a distance of 8 km due east. Draw the vector.

We first choose a suitable scale to represent the magnitude of the vector. In Fig. 9.2 a scale of 10 mm = 2 km has been chosen as convenient. We then draw a horizontal line 40 mm long and label the ends as shown. So $\overrightarrow{AB}$ represents the vector 8 km due east.

Sometimes we add an arrow to the vector (Fig. 9.2) to confirm the sense, but it is not necessary.

Cartesian components

Fig. 9.3 shows vector $\overrightarrow{AB}$ of magnitude 6 m/s and making an angle of 30° with the horizontal. An alternative is to define the vector AB by resolving it into its horizontal component $\overrightarrow{AC}$, and its vertical component $\overrightarrow{CB}$. These are said to be the Cartesian components of the vector.

$$\left.\begin{array}{c} \text{The magnitude of} \\ \text{component } \overrightarrow{AC} \end{array}\right\} = 6 \times \cos 30° = 5.20 \text{ m/s}$$

$$\left.\begin{array}{c} \text{The magnitude of} \\ \text{component } \overrightarrow{CB} \end{array}\right\} = 6 \times \sin 30° = 3 \text{ m/s}$$

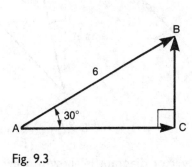

Fig. 9.3

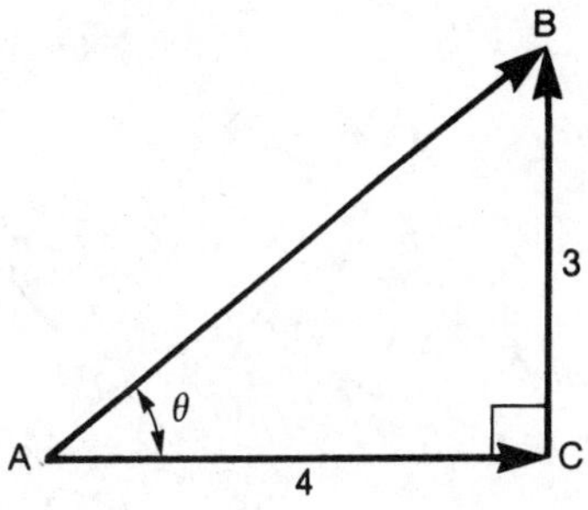

Fig. 9.4

EXAMPLE 9.2
The vector $\overrightarrow{AB}$ in Fig. 9.4 has a vertical component of 3 N and a horizontal component of 4 N. Calculate its magnitude and direction.

Using Pythagoras' theorem the magnitude of $\overrightarrow{AB} = \sqrt{4^2 + 3^3} = 5$ N.

To state its direction we find the size of angle θ.

$$\tan \theta = \tfrac{3}{4} \quad \text{giving} \quad \theta = 36.9°$$

Addition of vectors and resultant

Three points P, Q and R are marked out in a field, and are shown to scale in Fig. 9.5. A man walks from P to Q (i.e. he describes $\overrightarrow{PQ}$), and then walks on from Q to R (i.e. he describes $\overrightarrow{QR}$). Instead the man could have walked directly from P to R, thus describing the vector $\overrightarrow{PR}$.

Now going from P to R directly has the same result as going from P to Q and then from Q to R. We therefore call $\overrightarrow{PR}$ the *resultant* of the sum of the vectors $\overrightarrow{PQ}$ and $\overrightarrow{QR}$,
and we write this as

$$\overrightarrow{PR} = \overrightarrow{PQ} + \overrightarrow{QR}$$

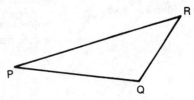

Fig. 9.5

EXAMPLE 9.3
Two forces act at a point O as shown in Fig. 9.6. Find the resultant force at O

a) by making a scale drawing,
b) by calculation.

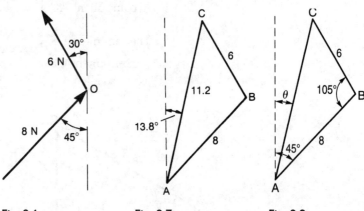

Fig. 9.6   **Fig. 9.7**   **Fig. 9.8**

a) The vector diagram is shown in Fig. 9.7. Draw AB parallel to the direction of the 8 N force, and of length representing 8 N to whatever scale you choose. Then draw BC parallel to the direction of the 6 N force and of length representing 6 N to scale. Here $\overrightarrow{AB} + \overrightarrow{BC} = \overrightarrow{AC}$, and thus $\overrightarrow{AC}$ represents the resultant of the two given forces. Measure the length AC and the angle it makes with the vertical.

Thus the resultant of the two forces is 11.2 N acting at 13.8° to the vertical.

b) Using the cosine rule $b^2 = a^2 + b^2 - 2ac \cos B$
   we have from $\triangle$ABC in Fig. 9.8:

$$AC^2 = 6^2 + 8^2 - 2 \times 6 \times 8 \times \cos 105°$$
$$= 124.8$$

giving $\qquad AC = 11.2 \, N$ confirming the drawing result.

Using the sine rule $\qquad \dfrac{a}{\sin A} = \dfrac{b}{\sin B}$

or $\qquad \sin A = \dfrac{a \sin B}{b}$

We have from $\triangle ABC$ in Fig. 9.8:

$$\sin A = \frac{6 \sin 105°}{11.2}$$

from which $\qquad A = $ inv sin 0.5175

$$= 31.2°$$

Hence $\qquad \theta = 45° - 31.2°$

$\qquad\qquad = 13.8°$ confirming the drawing result.

Vector addition using a parallelogram

In the previous example vector addition made use of a triangle. A parallelogram may also be considered as in Fig. 9.9.

We have $\qquad\qquad \overrightarrow{OA} + \overrightarrow{OB} = \overrightarrow{OR}$

Using the triangle OAR we would have said that $\overrightarrow{OA} + \overrightarrow{AR} = \overrightarrow{OR}$; this is a similar statement since $\overrightarrow{OB} = \overrightarrow{AR}$ because they are equal vectors having the same magnitude, direction and sense. Any numerical calculations would be carried out using $\triangle OAR$ as in Example 9.3.

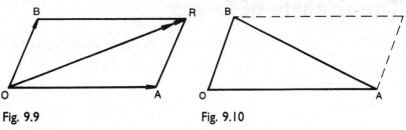

Fig. 9.9 $\qquad\qquad\qquad\qquad$ Fig. 9.10

Subtraction of vectors

The inverse of a vector is one having the same magnitude and direction, but of opposite sense. Thus $\overrightarrow{BA}$ is the inverse of $\overrightarrow{AB}$, or $\overrightarrow{BA} = -\overrightarrow{AB}$.

To subtract a vector we add its inverse. Hence if we wish to subtract vector $\overrightarrow{OB}$ from $\overrightarrow{OA}$ we have:

$$\overrightarrow{OA} - \overrightarrow{OB} = \overrightarrow{OA} + \overrightarrow{BO} = \overrightarrow{BO} + \overrightarrow{OA} = \overrightarrow{BA}$$

In Fig. 9.10 we can see that from the $\triangle OAB$ a vector sum gives $\overrightarrow{BO} + \overrightarrow{OA} = \overrightarrow{BA}$. Thus:

Vector difference $\qquad \overrightarrow{OA} - \overrightarrow{OB} = \overrightarrow{BA}$

This vector difference is represented by the other diagonal on the parallelogram than that used for the vector sum.

Phasors

In electrical engineering currents and voltages are represented by *phasors* in a similar manner to that in which vectors may be used to represent forces and velocities. The methods of adding and subtracting phasors and vectors are similar.

EXAMPLE 9.4

Phasor OA represents 7 V and phasor OB represents 5 V as shown in Fig. 9.11. Find the phasor difference $\overrightarrow{OA} - \overrightarrow{OB}$.

We suggest that you draw a diagram to scale and measure the result—this will serve as a check to the answer obtained by calculation.

Now $\overrightarrow{OA} - \overrightarrow{OB} = \overrightarrow{BA}$ which confirms that $\overrightarrow{BA}$ is the phasor difference. We are only concerned with lengths and angles in the calculations which follow.

Using the cosine rule for $\triangle OAB$ in Fig. 9.11:

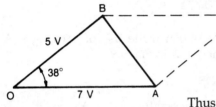

$$AB^2 = OA^2 + OB^2 - 2(OA)(OB) \cos \angle AOB$$
$$= 7^2 + 5^2 - 2 \times 7 \times 5 \times \cos 38°$$
$$= 18.84$$

from which $\quad AB = 4.34$

Fig. 9.11

Thus the phasor difference has a magnitude of 4.34 V and, if required, its direction may be found using the sine rule for $\triangle OAB$.

The triangle of forces

This is an important application of vector representation.

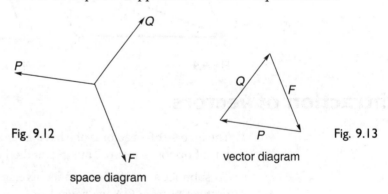

Fig. 9.12

space diagram

vector diagram

Fig. 9.13

If three forces act at a point (Fig. 9.12) and are in equilibrium, then they may be represented as vectors (Fig. 9.13) by the sides of a closed triangle.

The word 'closed' refers to the fact that the force vectors follow each other, nose to tail, all the way round the vector diagram with the nose of the last vector finishing up at the tail of the first. This idea may be extended to a polygon vector diagram where each side represents one force of many, in a system which is in equilibrium.

EXAMPLE 9.5

A mass of 7 kg is suspended from the ceiling as in Fig. 9.14. Find the tension forces in the supporting ropes a) by scale drawing and b) by calculation.

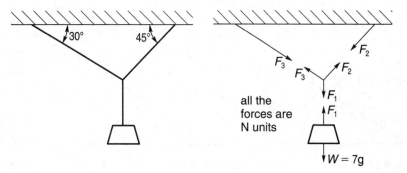

| Fig. 9.14 | Fig. 9.15 |

However complicated an engineering system is, it will become much easier if you get into the habit of redrawing the physical arrangement (Fig. 9.15), and splitting it into its component parts. This idea may be extended to frameworks where it simplifies the identification of tension or compression forces in the framework members.

Using either method of solution we shall be considering the equilibrium of the point where the ropes are joined. To find force F_1 we can turn our attention to the 7 kg mass. Since this mass is also in equilibrium then $F_1 = 7g$ N, which is the weight of the mass, and if we take $g = 10 \text{ ms}^{-2}$ then $F_1 = 70$ N.

a) Choose your own scale starting with the known 70 N force vector vertically downwards and draw the other two sides parallel to the lines of action of F_2 and F_3 respectively (see Fig. 9.16). Measurement will give you the values of the tension forces $F_2 = 63$ N and $F_3 = 51$ N to 2 s.f. (a reasonable accuracy for a graphical result).

b) Using the sine rule for the force vector triangle:

$$\frac{F_2}{\sin 60°} = \frac{70}{\sin 75°}$$

from which $\qquad F_2 = 62.8 \text{ N} \quad$ correct to 3 s.f.

and using the sine rule again to find F_3:

$$\frac{F_3}{\sin 45°} = \frac{70}{\sin 75°}$$

giving $\qquad F_3 = 51.2 \text{ N} \quad$ correct to 3 s.f.

These results confirm those obtained graphically.

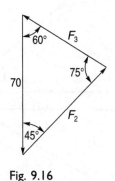

Fig. 9.16

EXAMPLE 9.6

A block of metal is at rest on an incline, as shown in Fig. 9.17. Find:

a) the friction force which is preventing the block from sliding down;

b) the normal reaction between the block and the plane.

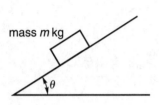

Fig. 9.17

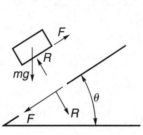

Fig. 9.18

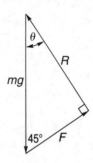

Fig. 9.19

In Fig. 9.18 we show the forces (in newtons) acting on the block itself, and their 'opposites' on the inclined plane (remember Newton's law 'to every action there is an equal and opposite reaction'). Although the forces do not strictly act at a point, the error is minimal and we draw the force vector triangle as in Fig. 9.19.

Since the vector triangle is right-angled then:

$$\sin \theta = \frac{\text{opposite side}}{\text{hypotenuse}}$$

from which   Friction force $F = mg \sin \theta$ newtons

Also $$\cos \theta = \frac{\text{adjacent side}}{\text{hypotenuse}}$$

giving   Normal reaction $R = mg \cos \theta$ newtons

Exercise 9.1

In Questions 1–4 find the values of the horizontal and vertical components illustrating your results in each case on a suitable vector diagram.

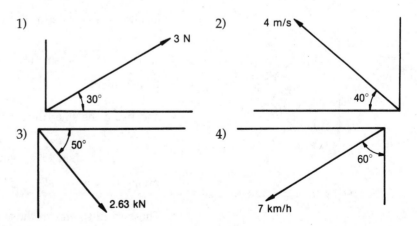

5) A block is being pulled up an incline, as shown in Fig. 9.20, using a rope making an angle of 25° with the incline. If the pull in the rope is 50 kN, draw a suitable vector diagram indicating the magnitudes of the components of this force parallel, and at right angles, to the incline.

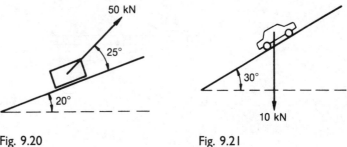

Fig. 9.20   Fig. 9.21

6) A motor car weighing 10 kN is shown on a 30° slope in Fig. 9.21. Resolve this weight into components along the slope and at right angles to it.

7) A garden roller is being pushed forwards with a force of 95 N as shown in Fig. 9.22. Find the horizontal and vertical components of this force.

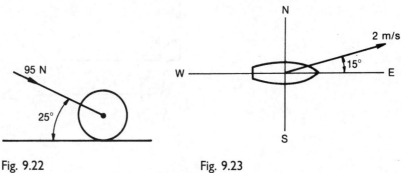

Fig. 9.22   Fig. 9.23

8) A ship is being steered in an easterly direction as shown in plan view in Fig. 9.23. However, owing to a S to N crosswind its motion and velocity are as shown. Find the forward speed of the vessel.

9) A sledge is being pulled on the level by two horizontal ropes which are at right angles to each other. If the respective tensions in the ropes are 30 N and 50 N, find the resultant force on the sledge and the angle it makes with the rope having the greater tension.

10) A plane is flying due north with a speed of 700 km/h. If there is a west to east cross-wind of 70 km/h, determine the resultant velocity of the plane and give its direction relative to north.

11) A dinghy is drifting under the influence of a 0.5 m/s tide flowing from north-east to south-west, and a wind of 0.8 m/s blowing from south to north. Find the resultant velocity of the dinghy and its direction relative to north.

12) The plan of a truck on rails is shown in Fig. 9.24. Find the resultant pull of the two ropes and the direction it acts in relative to the line of the rails.

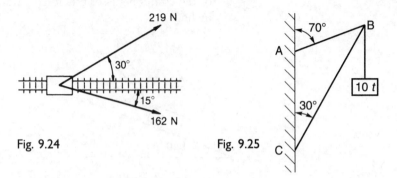

Fig. 9.24   Fig. 9.25

13) Using the results obtained to Question 12 find the force component pulling the truck along the track, and the force between the wheel flanges and the rails.

In Questions 14–17 find the resultants of the given phasors: in each case giving the direction relative to the larger given value.

14)   15)

16)   17)

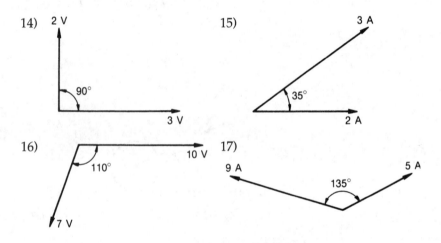

18) Find the forces in the members of the jib crane shown in Fig. 9.25. The units should be kN and you should take care to indicate if the member is in tension or compression.

Portfolio problems 2

1) A customer is experimenting with a mechanical feed system for animals and, because of its flexibility, has chosen an open belt drive for part of the power transmission train from the electric motor. The pulleys are 100 mm and 200 mm in diameter respectively, and their centres are 250 mm apart. In order to design and select a suitable belt he needs to know the smaller of the two angles of lap of the belt on the pulleys. This can then be used in the formula, which gives the ratio between the tension F_t N in the tight side and tension F_s N in the slack side of the belt drive, namely $F_t/F_s = e^{\mu\theta}$. In this expression μ is the coefficient of friction between the belt and the pulleys, and θ radians is the angle of lap. You have been asked to supply the following:

1. The total length of the belt assuming that there is no slack.
2. The smaller of the angles of lap.
3. The value of the tension in the slack side of the drive if the tension in the tight side is 250 N, assuming that 0.42 is the value of μ.

2)

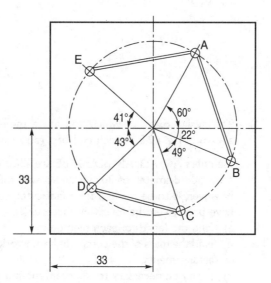

Fig. P2.1

A simple printed circuit is required for an electrical component. The designer involved tried to make the locations of the printed circuit 'terminals' A, B, C, D and E (Fig. P2.1) as simple as possible by putting them all on a pitch circle having a diameter of 54 mm, and giving angle references to the horizontal and vertical diameters of the pitch circle. Unfortunately one of the machines used in the manufacture of the circuit board could not be set up using polar coordinates from the pitch circle centre. This machine had to locate the circuit board on the left hand and bottom edges, and these were taken as the rectangular axes of reference. You have been asked by the production department to work out the rectangular coordinates of each terminal. And as if that isn't enough, they also need to know the lengths EAB and DC of the printed circuit.

3) One of the quality control inspectors has brought a machine shop measurement problem for you to solve—his excuse is that he has never been any good at arithmetic and that you are still at 'school'—you will get used to this sort of remark! However, he is a nice fellow and you decide to help. His problem is shown in Fig. P2.2. The use of rollers helps in the checking of the accuracy of machining operations. He requires the dimension *M*, the value of which will be used as a standard against measurements made during the production process.

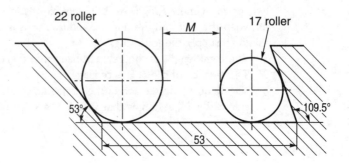

Fig. P2.2

4) The design team, of which you are a member, propose to use a framework as shown in Fig. P2.3 at each end of a gantry which supports infra-red drying lamps.

Members AC, AD, AB, BD and BE are all to be made from aluminium tubing having a outside diameter of 30 mm and a wall thickness of 5 mm. The horizontal base, however, is to be made from a fabricated steel section as shown in Fig. P2.4. You have been given the following three tasks:
a) Find the length of each member.
b) Find the mass of the complete framework, allowing an extra 10% for the jointing arrangements.
c) It may be necessary to 'sheet in' the framework at a later stage and in order to estimate the cost, you have been asked to work out the area ABEC.

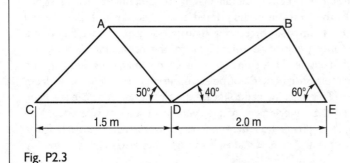

Fig. P2.3

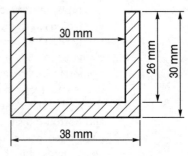

Fig. P2.4

5) A small company manufactures welded frameworks and has a design office in which you are working as a junior designer. A contract has been secured by the sales office for a triangular roof truss which will be loaded as shown in Fig. P2.5. You have been asked to find the forces in each of the three members of the truss, which will involve solving the triangle itself and then using the triangle of forces for the equilibrium of the three points A, B and C. Two of these are absolutely necessary and the third will act as a check. The customer is a 'belt and braces' individual who has requested that your report should be accompanied by scale drawings of the framework and also the force diagrams as a check on your theoretical calculations.

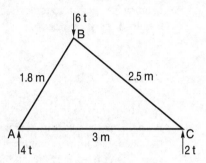

Fig. P2.5

Portfolio assignment 2

In the five chapters of this part we have looked at the use of trigonometry. We have introduced a number of methods to find lengths, angles and areas of various triangles. The book has shown how these mathematical methods can be applied to solve the problems of engineers, in both the chapter and portfolio exercises. However, the main skill is knowing which of the mathematical methods to use for which problem. This assignment will give you the opportunity to use this skill to formulate a solution to a complex problem using a number of techniques. This is the type of evidence which is required for your portfolio.

A production line

Production brief

A company manufacturing frozen ice drinks uses conveyor belts to transport the filled, but unfrozen, containers through each stage of production. At one stage the conveyor is inclined 35° from the horizontal to allow the drinks to enter a cooler (Fig. P2.6). We can assume that no slipping will occur as a ribbed belt will be used.

The production manager is concerned that the new larger drinks which are to be introduced may be too tall to travel up the incline and will become unstable and topple over, spilling the drink on the conveyor before freezing. As his assistant, he wants you to check that his theory is correct and to recalculate the maximum angle of incline to which the conveyor may be set, so that the drinks do not topple down the conveyor. You are told that the width of the new containers is to remain the same, but the height from the base to the centre of gravity is increased by 20%.

Use the formula $\tan \Phi = \dfrac{1/2w}{b}$.

where w = width of the base
b = height of the centre of gravity from the base
Φ = angle of incline

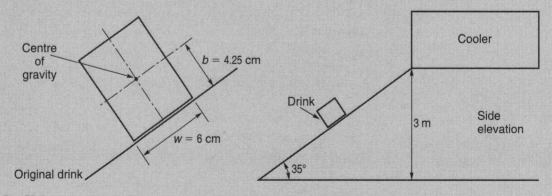

Fig. P2.6

As there is a shortage of floor space, it is also necessary to calculate how much longer the horizontal distance of the new conveyor needs to be to reach the height of the cooler.

To aid production, reduce the monotony of the workforce and improve health and safety conditions in the packing of these frozen drinks, a pick and place robot is to be installed. The robot will move the drinks from the cooler to the packaging production line of the factory.

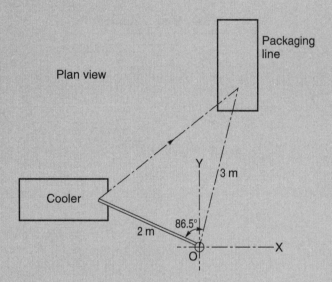

Fig. P2.7

The robot arm extends from 2 m to 3 m while turning through an angle of 86.5°. As the robot arm works on a continuous path system (the arm end will move in a straight line) the production team need to know, for programming purposes, the polar coordinate positions of the pick up and put down points of the drinks, taking the centre of the robot as a datum point (O), see Fig. P2.7.

The health and safety team need to know for factory plans the area that the robotic arm sweeps through.

Task 1
a)   Calculate the height of the centre of gravity of the new drink.
b)   Calculate the angle of the new conveyor.
c)   Find how much longer the new conveyor will need to be if the drinks are still to load into the cooler at a height of 3 m.

Task 2
a)   Find the polar coordinate positions of the put down and the pick up points of the robotic arm assuming the datum for your calculations to be lines OX and OY.
b)   Calculate the sweep area of the robotic arm.

Unit coverage

Table E2 below shows which parts of the unit have been covered in this assignment and where there are opportunities to assess core skills. If mapping of this type is shown for all your portfolio work, it is easy to check that you have covered all the elements in the unit.

Table E2

Task	Unit	Element/PC	Core skills
1	8	2.1, 2.2, 2.3, 2.4	Application of number 3.1.2 and 3.2.3
2	8	2.2, 2.3	Application of number 3.2.1 and 3.2.6

PART THREE

Use of functions and graphs to model engineering situations and solve engineering problems

By now you will realise how important algebraic functions, such as equations, are as tools for the solution of engineering problems. Many functions may be illustrated graphically and this gives us another approach to analysing data and the resulting calculations.

Motor vehicle engineers use an oscilloscope which displays the 'firing lines', which is a very useful graphical display of the ignition voltage in each cylinder of an engine.

Electrical engineers find the portrayal of trigonometrical waveforms extremely effective in the analysis of alternating electrical currents and voltages.

Some equation relationships, or laws as they are sometimes called, may be rearranged to give a simpler linear display, that is a straight line graph. Again a graphical approach reduces a complicated arrangement to a much simpler form.

Differentiation and integration are more 'tools for the engineer'. The terms sound frightening but, after a few hours of study, you will realise that the techniques are no more difficult to understand than anything else. We can use these ideas of differential calculus to examine problems from topics such as velocity, acceleration, small errors and even the optimum shape for a 'mushy peas' can!

Go ahead and enjoy drawing and sketching graphs — it gives you a break from algebra!

Functions and graphs

In newspapers, television, business reports, technical papers and government publications, use is made of pictorial illustrations to compare quantities. These diagrams help the reader to understand what deductions may be drawn from the quantities represented. The most common sort of diagram is the graph. An everyday example would be industrial output against time in months. But for us, as engineers, a more likely example would be acceleration of a vehicle against time in seconds.

Graphs of linear equations

Linear equations produce straight line graphs—these are very useful as they are straightforward to draw and their equations are simple when compared with others. Here you will learn how to choose axes and appropriate scales, draw a line which fits the plotted points and then calculate the relevant results.

Axes of reference and coordinates

The graphs which follow will all be based on Cartesian or rectangular axes—we have met these already in Chapter 5. We shall also use rectangular coordinates to plot the individual points from which the graphs will be drawn.

Axes and scales

The location of the axes on the graph paper, and the choice of scale along each axis is largely a matter of experience. The more practice you have in dealing with two sets of correspnding data and then having to choose.

The position of axes and the scales on each axis (which need not be the same), plot points and draw the graph, the easier it will be.

You may appreciate some advice on the choice of a scale on an axis. The scale should be as large as possible so that the points may be plotted with the greatest possible accuracy—this means that you should make the plot cover the whole sheet of graph paper, not just a small area in the bottom left-hand corner!

The difference between a good scale and a bad scale is how easy it is to read intermediate decimal values—try 1.7 on the 'bad' scales shown below and you will see why they are so named.

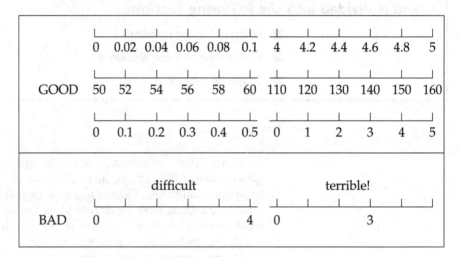

Interpolation and extrapolation

The following examples illustrate these terms.

EXAMPLE 10.1

The table below gives the corresponding values of x and y. Plot this information and from the graph find:

a) the value of y when $x = -3$

b) the value of x when $y = 2$

x	−4	−2	0	2	4	6
y	−2.0	−1.6	0	1.4	2.5	3.0

The graph is shown plotted in Fig. 10.1 and it is a smooth curve. This means that there is a definite law (or equation) connecting x and y. We can therefore use the graph to find corresponding values of x and y

between those given in the original table of values. By using the constructions shown in Fig. 10.1

a) the value of y is -1.9 when $x = -3$.

b) the value of x is 3 when $y = 2$

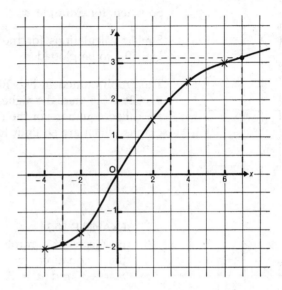

Fig. 10.1

Using a graph in this way to find values of x and y not given in the original table of values is called *interpolation*. If we extend the curve so that it follows the general trend we can estimate corresponding values of x and y which lie *just beyond* the range of the given value. Thus in Fig. 10.1 by extending the curve we can find the probable value of y when $x = 7$. This is found to be 3.2.

Finding a probable value in this way is called *extrapolation*. An extrapolated value can usually be relied upon but in some cases it may contain a substantial amount of error. Extrapolated values must therefore be used with care.

It must be clearly understood that interpolation and extrapolation can only be used if the graph is a straight line or a smooth curve.

Graphs of simple equations

Consider the equation: $y = 2x + 5$
We can give x any value we please and so calculate a corresponding value for y. Thus, when $x = 0$ $y = 2 \times 0 + 5 = 5$

                  when $x = 1$ $y = 2 \times 1 + 5 = 7$

                  when $x = 2$ $y = 2 \times 2 + 5 = 9$ and so on.

The value of y therefore depends on the value allocated to x. We therefore call y the *dependent variable*. Since we can give x any value we please, we call x the *independent variable*. It is usual to mark the values of the independent variable along the horizontal x-axis and the values of the dependent variable are then marked off along the vertical y-axis.

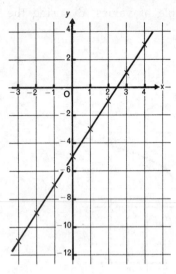

Fig. 10.2

EXAMPLE 10.2

Draw the graph of $y = 2x - 5$ for values of x between -3 and 4.

We must first decide on which x values to use—here the choice is reasonably easy, say, -3, -2, -1, 0, 1, 2 and 3.

So when, for example $x = -2$, then $y = 2(-2) - 5 = -9$.

Similar calculations for the other values enable the graph shown in Fig. 10.2 to be plotted.

The graph plotted in Fig. 10.2 is a straight line. Equations of the type $y = 2x - 5$, where the highest power of the variables x and y is the first, are called equations of the first degree. All equations of this type give graphs which are straight lines and hence they are often called linear equations.

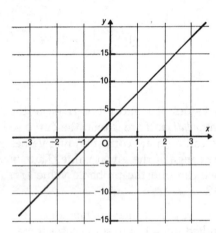

Fig. 10.3

EXAMPLE 10.3

By means of a graph show the relationship between x and y in the equation $y = 5x + 3$. Plot the graph between $x = -3$ and $x = 3$.

Although we can draw a straight line graph using only two points, it is usual to take three points, the third acting as a check on the other two.

x	-3	0	$+3$
$y = 5x + 3$	-12	3	$+18$

The graph is shown in Fig. 10.3.

The straight line

As we have seen already a straight line results from plotting a linear equation, which is an equation of the first degree.

A straight line graph is probably the commonest form of graph and this is why we study it first. It is even more important because we can often reduce more complicated equations to a straight line form.

The law of a straight line

In Fig. 10.4, the point B is any point on the line shown and has coordinates x and y. Point A is where the line cuts the y-axis and has coordinates $x = 0$ and $y = c$.

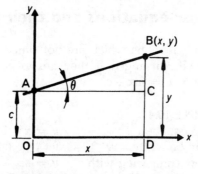

In $\triangle ABC$

$$\frac{BC}{AC} = \tan\theta$$

$$\therefore \quad BC = (\tan\theta) \cdot AC$$

but also

$$y = BC + CD$$
$$= (\tan\theta)\,AC + CD$$
$$= mx + c$$

Fig. 10.4

$$\therefore \qquad y = mx + c$$

This is called the *standard equation*, or *law*, of a straight line

c is called the intercept on the y-axis
Care must be taken as this only applies if the zero of the scale along the x-axis is at the intersection of the x and y axes.

m is called the gradient
In mathematics, the gradient* of a line is defined as the tangent of the angle θ.

Hence in Fig. 10.4 the gradient $m = \tan\theta = \dfrac{BC}{AC}$

Positive and negative gradients are shown in Fig. 10.5.

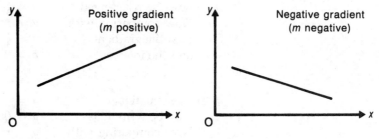

Fig. 10.5

Two special cases—zero gradient and infinite gradient—are shown in Fig. 10.6.

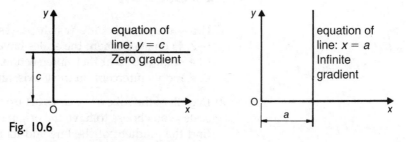

Fig. 10.6

---

*Care should be taken not to confuse this with the gradient given on maps, railways, etc. which is the sine of the angle (not the tangent)—e.g. a railway slope of 1 in 100 is one unit vertically for every 100 units measured along the slope.

Linear equations and their graphs

Linear equations which are not stated in standard straight line form must be rearranged if they are to be compared with the standard equation $y = mx + c$.

EXAMPLE 10.4

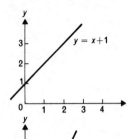

a) The equation $\qquad y = x + 1$
   may be written as $\qquad y = 1x + 1$
   Now comparing with $\qquad y = mx + c$
   then the gradient $\qquad m = 1$
   and the intercept $\qquad c = 1$

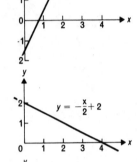

b) The equation $\qquad y = 2x - 1.5$
   may be compared with $\quad y = mx + c$
   then the gradient $\qquad m = 2$
   and the intercept $\qquad c = -1.5$

c) The equation $\qquad 2y + x = 4$
   must be arranged: $\qquad y = -\frac{1}{2}x + 2$
   Now comparing with $\qquad y = mx + c$
   then the gradient $\qquad m = -\frac{1}{2}$
   and the intercept $\qquad c = 2$

d) The equation $\qquad 3y = x$
   must be rearranged: $\qquad y = \frac{x}{3} + 0$
   Now comparing with $\qquad y = mx + c$
   then the gradient $\qquad m = \frac{1}{3}$
   and the intercept $\qquad c = 0$

e) The equation $\qquad y = -2$
   may be written as $\qquad y = 0x - 2$
   Now comparing with $\qquad y = mx + c$
   then the gradient $\qquad m = 0$
   and the intercept $\qquad c = -2$

Exercise 10.1

1) Draw the straight line which passes through the points (4, 7) and (−2, 1) on axes where the scales have been chosen so that the origin (the point 0, 0) is at their intersection. Hence, find the gradient of the line and its intercept on the y-axis, and state the equation of the line.

2) In each of the following cases set up x and y axes and arrange for the scales you choose to have the origin at their intersection. In each case find the gradient of the line and its intercept on the y-axis:

   a) $y = x + 3$ b) $y = -3x + 4$ c) $y = -3.1x - 1.7$
   d) $y = 4.3x - 2.5$ e) $x = -1$ f) $y = 2.9$
   g) $x = y + 4$ h) $x + 2y = 3$ j) $2x/y = 3.6$

Obtaining the straight line law of a graph

Two methods are used:

1. Origin at the intersection of the axes

When it is convenient to arrange the origin, i.e. the point $(0, 0)$, at the intersection of the axes the values of gradient m and intercept c may be found directly from the graph as shown in Example 10.5.

EXAMPLE 10.5
Find the law of the straight line shown in Fig. 10.7.

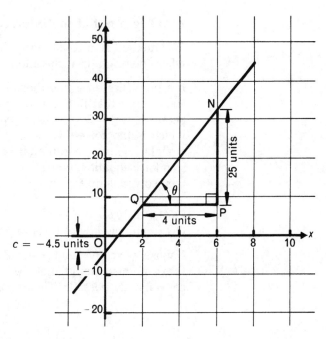

Fig. 10.7

To find gradient m.
Take any two points Q and N on the line and construct the right-angled triangle QPN. This triangle should be of reasonable size, since a small triangle will probably give an inaccurate result. Note that if we can measure to an accuracy of 1 mm using an ordinary rule, then this error in a length of 20 mm is much more significant than the same error in a length of 50 mm.

The lengths of NP and QP are then found using the scales of the x and y axes. Direct lengths of these lines, as would be obtained using an ordinary rule, e.g. both in centimetres, must *not* be used—the scales of the axes must be taken into account.

$$\therefore \qquad \text{Gradient } m = \tan \theta = \frac{\text{NP}}{\text{QP}} = \frac{25}{4} = 6.25$$

To find intercept c.

This is measured again using the scale of the y-axis.

$\therefore$ intercept $\qquad\qquad\qquad c = -4.5$

The law of the straight line

The standard equation is

$$y = mx + c$$

$\therefore$ the required equation is

$$y = 6.25x + (-4.5)$$

i.e. $\qquad\qquad\qquad y = 6.25x - 4.5$

2. Origin not at the intersection of the axes

This method is applicable for all problems—it may be used, therefore, when the origin is at the intersection of the axes.

If a point lies on a line then the coordinates of that point satisfy the equation of that line, e.g. the point (2, 7) lies on the line $y = 2x + 3$ because if $x = 2$ is substituted in the equation, $y = 2 \times 2 + 3 = 7$ which is the correct value of y. Two points, which lie on the given straight line, are chosen and their coordinates are substituted into the standard equation $y = mx + c$. The two equations which result are then solved simultaneously to find the values of m and c.

EXAMPLE 10.6

Determine the law of the straight line shown in Fig. 10.8.

Choose two convenient points P and Q and find their cordinates. Again these points should not be close together, but as far apart as conveniently possible. Their coordinates are as shown in Fig. 10.8.

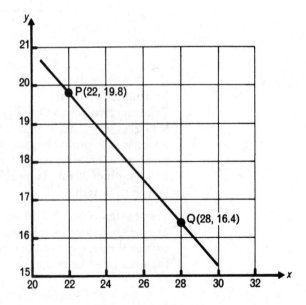

Fig. 10.8

204

Let the equation of the line be $y = mx + c$.

Now P (22, 19.8) lies on the line $\therefore$ $19.8 = m(22) + c$

and Q (28, 16.4) lies on the line $\therefore$ $16.4 = m(28) + c$

To solve these two equations simultaneously we must first eliminate one of the unknowns. In this case c will disappear if the second equation is subtracted from the first, giving

$$19.8 - 16.4 = m(22 - 28)$$

$$\therefore \qquad m = \frac{3.4}{-6} = -0.567$$

To find c the value of $m = -0.567$ may be substituted into either of the original equations. Choosing the first equation we get

$$19.8 = -0.567(22) + c$$

$$\therefore \qquad c = 19.8 + 0.567(22) = 32.3$$

Hence the required law of the straight line is

$$y = -0.567x + 32.3$$

Exercise 10.2

1) A straight line passes through the points $(-2, -3)$ and $(3, 7)$. *Without drawing the line* find the values of m and c in the equation $y = mx + c$.

2) The width of keyways for various shaft diameters are given in the table below.

Diameter of shaft D (mm)	10	20	30	40	50	60	70	80
Width of key-way W (mm)	3.75	6.25	8.75	11.25	13.75	16.25	18.75	21.25

Show that D and W are connected by a law of the type $W = aD + b$ and find the values of a and b.

3) During an experiment to find the coefficient of friction between two metallic surfaces the following results were obtained.

Load W (N)	10	20	30	40	50	60	70
Friction force F (N)	1.5	4.3	7.6	10.4	13.5	15.6	18.8

Show that F and W are connected by a law of the type $F = aW + b$ and find the values of a and b.

4) In a test on a certain lifting machine it is found that an effort of 50 N will lift a load of 324 N and that an effort of 70 N will lift a load of 415 N. Assuming that the graph of effort plotted against load is a straight line, find the probable load that will be lifted by an effort of 95 N.

Graphs of experimental data

Best fit straight line

Readings which are obtained as a result of an experiment will usually contain errors owing to inaccurate measurement and other experimental errors. If the points, when plotted, show a trend towards a straight line or a smooth curve, this is usually accepted and the best fit straight line or curve drawn. In this case the line will not pass through some of the points and an attempt must be made to ensure an even spread of these points above and below the line or the curve.

One of the most important applications of the straight line is the determination of a law connecting two quantities when values have been obtained from an experiment as illustrated by the following examples.

EXAMPLE 10.7

Hooke's law states that for an elastic material, up to the limit of proportionality, the stress, σ, is directly proportional to the strain, ε, it produces. In equation form this is $\sigma = E\varepsilon$ where the constant E is called the modulus of elasticity of the material.

Find the value of E using the following results obtained in an experiment.

σ MN/m^2	125	110	95	80
ε no units	0.000 600	0.000 522	0.000 466	0.000 382

σ MN/m^2	63	54	38	
ε no units	0.000 367	0.000 269	0.000 168	

In order to compare the equation $\sigma = E\varepsilon$ with the standard straight line equation $y = mx + c$ we must plot σ on the vertical axis, and ε on the horizontal axis.

Inspection of the values shows that it is convenient to arrange for the origin, i.e. the point $(0, 0)$, to be at the intersection of the axes. The graph is shown in Fig. 10.9.

Since the values are the result of an experiment it is unlikely that the points plotted will lie exactly on a straight line. Having arranged for the origin to be at the intersection of the axes we can see from the given equation that the intercept on the y-axis is zero. This means that the graph passes through the origin and this helps us in drawing the 'best' straight line to 'fit' the points. Judgement is needed here—for instance the point $(0.000\,367, 63)$ is obviously an incorrect result as it lies well away from the line of the other points, and so we should ignore this point.

We see also that the gradient $m = E$ on our graph. Thus we may find the value of E by calculating the gradient of the straight line. From the suitable right-angled triangle POM

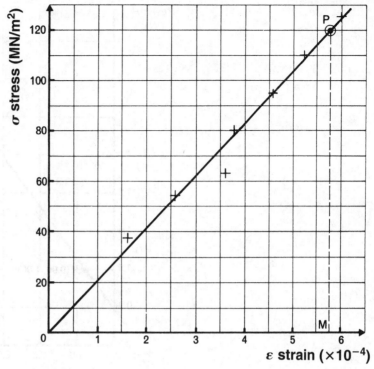

Fig. 10.9

the gradient is given by $\dfrac{\text{PM}}{\text{OM}} = \dfrac{120}{0.000\ 58} = 207\ 000$

The units of the ratio $\dfrac{\text{PM}}{\text{OM}}$ will be those of PM, i.e. MN/m^2, since OM represents strain which has no units (this is becase strain is the ratio of two lengths).

Hence the value of the modulus of elasticity of the material $E = 207\ 000\ \text{MN/m}^2 = 207\ \text{GN/m}^2$.

EXAMPLE 10.8
During a test to find how the power of a lathe varied with the depth of cut, results were obtained as shown in the table. The speed and feed of the lathe were kept constant during the test.

Depth of cut, d (mm)	0.51	1.02	1.52	2.03	2.54	3.0
Power, P (W)	0.89	1.04	1.14	1.32	1.43	1.55

Show that the law connecting d and P is of the form $P = ad + b$ and find the law. Hence find the value of d when P is 1.2 watts.

The standard equation of a straight line is $y = mx + c$. It often happens that the variables are *not* x and y. In this example d is used instead of x and is plotted on the horizontal axis, and P is used instead of y and is plotted on the vertical axis.

Similarly the gradient $= a$ instead of m,
and the intercept on the y-axis $= b$ instead of c.

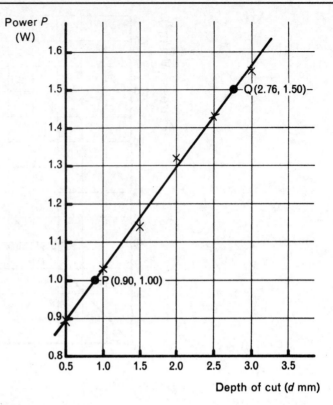

Fig. 10.10

On plotting the points (Fig. 10.10) it will be noticed that they deviate slightly from a straight line. Since the data are experimental we must expect errors in observation and measurement and hence a slight deviation from a straight line must be expected.

The points, therefore, approximately follow a straight line and we can say that the equation connecting P and d is of the form $P = ad + b$.

Because the origin is *not* at the intersection of the axes, to find the values of constants a and b we must choose two points *which lie on the line*. These two points must be as far apart as possible in order to obtain maximum accuracy.

In Fig. 10.10 the points P(0.90, 1.00) and Q(2.76, 1.50) have been chosen.

The point P(0.90, 1.00) lies on the line $\quad\therefore\quad 1.00 = a(0.90) + b$
The point Q(2.76, 1.50) lies on the line $\quad\therefore\quad 1.50 = a(2.76) + b$

Now subtracting the first equation from the second we get

$$1.50 - 1.00 = a(2.76 - 0.90)$$

$$\therefore \qquad a = \frac{0.50}{1.86} = 0.27$$

Now substituting the value $a = 0.27$ into the first equation we get

$$1.00 = 0.27(0.90) + b$$

$$\therefore \qquad b = 1.00 - 0.27(0.90) = 0.76$$

Hence the required law of the line is $P = 0.27d + 0.76$

To find d when $P = 1.2\,\text{W}$. The value $P = 1.2$ is substituted into the equation giving

$$1.2 = 0.27d + 0.76$$

$$d = \frac{1.2 - 0.76}{0.27} = 1.63\ \text{mm}$$

(since all values of d are in millimetres when values of P are in watts).

This value of d may be verified by checking the appropriate value of d corresponding to $P = 1.2$ on the straight line in Fig. 10.10.

Any inaccuracies may be due to rounding off calculations to two significant figures, e.g. the value of m is $\dfrac{0.5}{1.86} = 0.269$ if three significant figures are considered. Bearing in mind, however, the experimental errors etc. the rounding off as shown seems reasonable. This question of accuracy is always open to debate, the most dangerous error being to give the calculated results to a far greater accuracy than the original given data of warrants.

Exercise 10.3

1) The following results were obtained from an experiment on a set of pulleys. W is the load raised and E is the effort applied. Plot these results and obtain the law connecting E and W.

W(N)	15	20	25	30	35	40	45
E(N)	2.3	2.7	3.2	3.8	4.3	4.7	5.3

2) During a test with a thermocouple pyrometer the e.m.f. (E millivolts) was measured against the temperature at the hot junction ($t°C$) and the following results were obtained:

t	200	300	400	500	600	700	800	900	1000
E	6	9.1	12.0	14.8	18.2	21.0	24.1	26.8	30.2

The law connecting t and E is supposed to be $E = at + b$. Test if this is so and find suitable values for a and b.

3) The resistance (R ohms) of a field winding is measured at various temperatures ($t°C$) and the results are recorded in the table below:

$t\,(°C)$	21	26	33	38	47	54	59	66	75
R (ohms)	109	111	114	116	120	123	125	128	132

If the law connecting R and t is of the form $R = a + bt$ find suitable values of a and b.

4) The rate of a spring, λ, is defined as force per unit extension. Hence for a load, F N producing an extension x mm the law is $F = \lambda x$. Find the value of λ in units of N/m using the following values obtained from an experiment:

F (N)	20	40	60	80	100	120	140
x (mm)	37	79	111	156	197	229	270

5) A circuit contains a resistor having a fixed resistance of R ohms. The current, I amperes, and the potential difference, V volts, are related by the expression $V = IR$. Find the value of R given the following results obtained from an experiment:

V (volts)	3	7	11	13	17	20	24	29
I (amperes)	0.066	0.125	0.209	0.270	0.324	0.418	0.495	0.571

Graphs of non-linear equations

The recognition of the shapes and layout of curves related to their equations is important to us as technologists, although, at this stage of your studies, close examination of the properties of many curves is not possible due to time constraints. But we will look at the parabola, which is often used for the shape of headlamp reflectors, since a bulb placed at the focus will produce reflected rays which are parallel. Exponent curves, where x is replaced by t for time, illustrate growth and decay which are of particular interest to electrical engineers in the behaviour of electrical charges, currents and voltages.

Graphs of a familiar equation — the parabola

Here we will look at the various forms taken by a parabola when plotted as the graph of a quadratic equation $y = ax^2 + bx + c$.

The important part of the curve is usually the portion in the vicinity of the vertex of the parabola (Fig. 10.11).

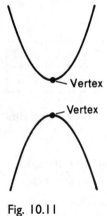

The shape and layout will depend on the values of the constants a, b and c and we will examine the effect of each constant in turn.

Fig. 10.11

Constant a

Consider the equation $y = ax^2$. The table of values given below is for $a = 4$, $a = 2$ and $a = -1$:

x	-3	-2	-1	0	1	2	3
$y = 4x^2$	36	16	4	0	4	16	36
$y = 2x^2$	18	8	2	0	2	8	18
$y = -x^2$	-9	-4	-1	0	-1	-4	-9

Fig. 10.12 shows the graphs of $y = ax^2$ when $a = 4, a = 2$ and $a = -1$.

We can see that if the value of a is positive the curve is shaped $\smile$, and the greater the value of a, the 'steeper' the curve rises.

Negative values of a give a curve shaped $\frown$.

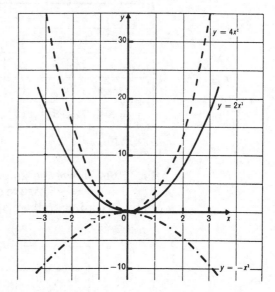

Fig. 10.12

Constant b

Consider the equation $y = x^2 + bx$. The table of values given below is for $b = 2$ and $b = -3$.

x	-3	-2	-1	0	1	2	3	4
$y = x^2 + 2x$	3	0	-1	0	3	8	15	24
$y = x^2 - 3x$	18	10	4	0	-2	-2	0	4

Fig. 10.13 shows the graphs of $y = x^2 + bx$ when $b = 2$ and $b = -3$. The effect of a positive value of b is to move the vertex to the left of the vertical y-axis, whilst a negative value of b moves the vertex to the right of the vertical axis.

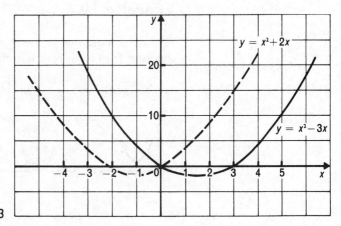

Fig. 10.13

Constant c

Consider the equation $y = x^2 - 2x + c$. The table of values given below is for $c = 10$, $c = 5$, $c = 0$ and $c = -5$:

x	-3	-2	-1	0	1	2	3	4
$y = x^2 - 2x + 10$	25	18	13	10	9	10	13	18
$y = x^2 - 2x + 5$	20	13	8	5	4	5	8	13
$y = x^2 - 2x$	15	8	3	0	-1	0	3	8
$y = x^2 - 2x - 5$	10	3	-2	-5	-6	-5	-2	3

Fig. 10.14 shows the graphs of $y = x^2 - 2x + c$ when $c = 10$, $c = 5$ $c = 0$ and $c = -5$. As we can see the effect is to move the vertex up or down according to the magnitude of c.

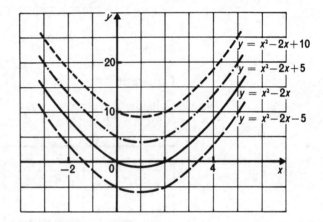

Fig. 10.14

Exercise 10.4

State which answer or answers are correct in Questions 1–6. In every diagram the origin is at the intersection of the axes.

1) The graph of $y = x^2$ is:

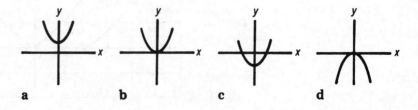

2) The graph of $y = x^2 + 2$ is:

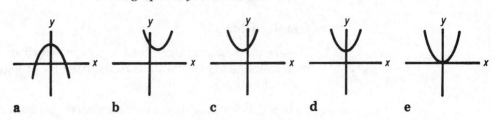

a   b   c   d   e

3) The graph of $y = -2x^2$ is:

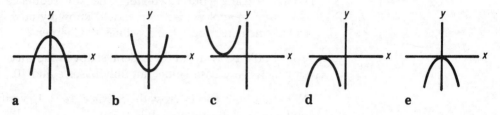

a   b   c   d   e

4) The graph of $y = 4 - x^2$ is:

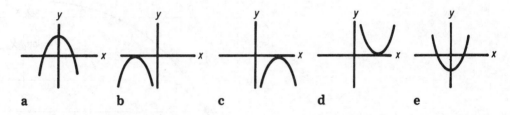

a   b   c   d   e

5) The graph of $y = x^2 - 3x + 2$ is:

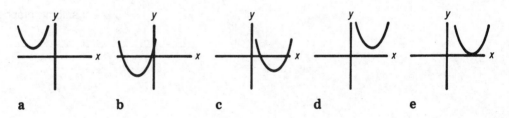

a   b   c   d   e

6) The graph of $y = x^2 + 4x + 4$ is:

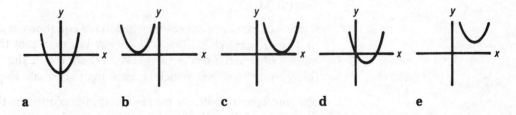

a   b   c   d   e

7) The profile of the cross-section of a headlamp reflector is parabolic in shape, having an equation $y = 0.00617x^2$. Plot the profile for values of x from 0 to 90 mm.

The logarithmic curve

Graph of log x

The graphs of natural logarithms, $\log_e$ or ln, which are logarithms to the base e, and also of logarithms to the base 10, or $\log_{10}$ or lg, are both shown plotting in Fig. 10.15.

Only the 'interesting' portion of each curve has been drawn, as we can easily judge what will happen at the extremes.

You may like to check some of the values, or perhaps plot the curves yourself, using your calculator. The key sequence varies from one calculator to another, but you should know by now how to use yours, as we calculated logarithmic values in Chapter 2.

You should be aware of the general shape of the curves which would be similar for any base value, not only bases e and 10.

Note that both curves cross the x-axis at $x = 1$, verifying the fact that both $\log_e 1 = 0$ and $\log_{10} 1 = 0$.

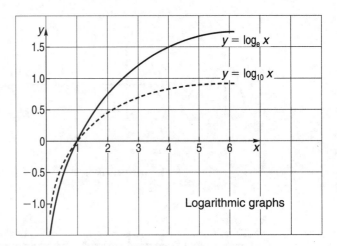

Fig. 10.15

The exponent curve

Graph of e^x

Curves of exponent functions which have equations of the type $y = e^x$ or $y = e^{-x}$ are called *exponent graphs*. We may plot the curves using values obtained from a calculator. When $x = 2$ the value of e^{-2} is 0.135 correct to 3 s.f., which is more than adequate to plot the graph.

For convenience both the curves are shown plotted on the same axes in Fig. 10.16. Although the range of values chosen for x is limited the overall shape of the curves is clearly shown.

The rate at which a curve is changing at any point is given by the gradient of the tangent at that point.

Remember the sign convention for gradients is

Positive gradient   Negative gradient

Now the gradient at any point on the e^x graph is positive and so the rate of change is positive. In addition the rate of change increases as the values of x increase. A graph of this type is called a *growth curve*.

The gradient at any point on the e^{-x} graph is negative and so the rate of change is negative. In addition the rate of change decreases as the values of x decrease. A graph of this type is called a *decay curve*.

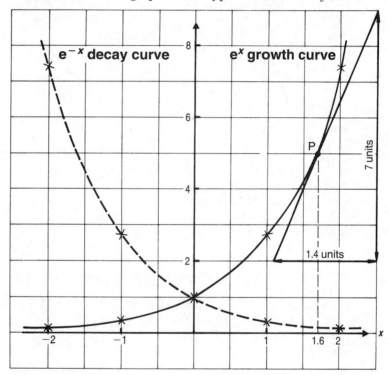

Fig. 10.16

An important property of the exponent function, e^x

Suppose we choose any point, P, on the graph of the function e^x shown in Fig. 10.16 and draw a tangent to the curve at P. We may find the gradient of the curve at P by finding the gradient of the tangent using the right-angled triangle shown. The gradient at $P = \dfrac{7}{1.4} = 5$. Now this is also the value of e^x at P where $x \approx 1.6$ giving $e^{1.6} \approx 5$.

You may like to draw the curve of e^x and check that the gradient at various points is always equal to the value of e^x at corresponding points. This illustrates the important, and unique, property of the exponent function e^x which is:

> At any point on the curve of the exponent function e^x the gradient of the curve is equal to the value of the function.

Using a time base — the curve of e^t

In technology reference to the exponent curves is usually based on time. Thus the curves are plotted on a time (t) base and not an x base.

Practical applications based on growth and decay start from a given instant. This means we are dealing with actual (or real) times which, in numerical terms, are positive values of t. Thus only portions of the curves to the right of the vertical axis are generally used, as shown in Fig. 10.17.

Another form of growth has an equation of the form $y = 1 - e^{-t}$ and this curve is shown in Fig. 10.18.

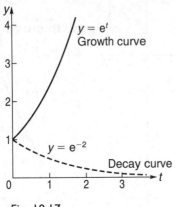

Fig. 10.17

Fig. 10.18

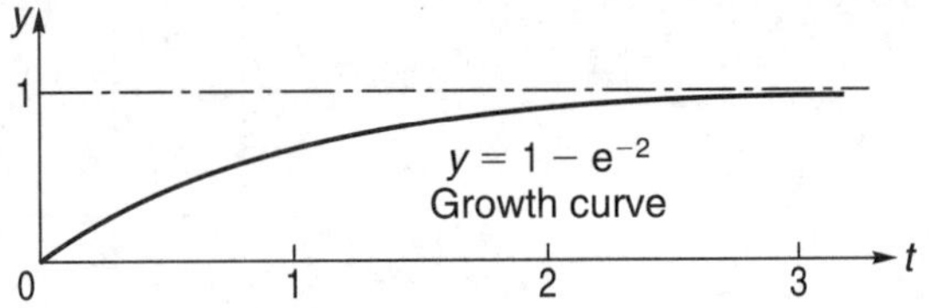

General expressions of growth and decay

When used to solve problems the equations are modified to the following general forms

$$y = ae^{bt}, \quad y = ae^{-bt}, \quad \text{and} \quad y = a(1 - e^{-bt})$$

where a abd b are constants which affect the initial value of y and also the rate of growth or decay. Sketches of the curves and some typical applications of each are shown in Figs 10.19, 10.20 and 10.21.

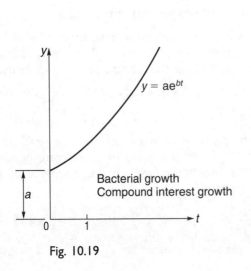

Fig. 10.19

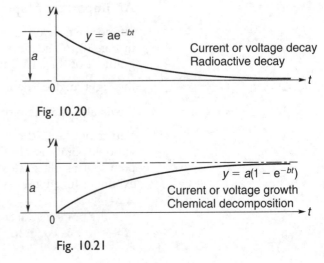

Fig. 10.20

Fig. 10.21

EXAMPLE 10.9

The instantaneous e.m.f. in an inductive circuit is given by the expression $100e^{-4t}$ volts, where t is time in seconds. Plot the graph of the e.m.f. for values of t from 0 to 0.5 seconds, and use the graph to find:

a) the value of the e.m.f. when $t = 0.25$ seconds, and
b) the rate of change of the e.m.f. when $t = 0.1$ seconds.

The graph is shown plotted in Fig. 10.22, from values obtained from a calculator.

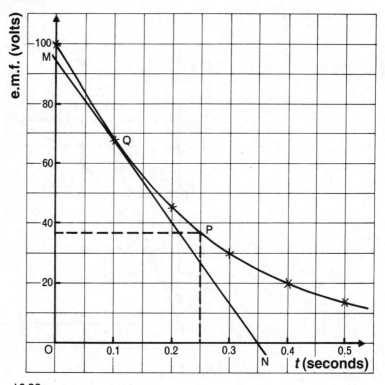

Fig. 10.22

a) The point P on the curve is at 0.25 seconds shown on the t scale and the corresponding value of e.m.f. can be read directly from the vertical axis scale. The value is 37 volts.

b) The point Q on the graph is at 0.1 seconds. Now the rate of change of the curve at Q is given by the gradient of the tangent at Q. This gradient may be found by constructing a suitable right-angled triangle such as MNO in Fig. 10.22 and finding the ratio $\dfrac{MO}{ON}$. Hence the

$$\text{Gradient at Q} = \frac{MO}{ON} = \frac{94 \text{ volts}}{0.35 \text{ seconds}} = 269 \text{ volts per second}$$

According to the sign convention a line sloping downwards from left to right has a negative gradient.

Hence the gradient at Q is −269 volts per second, which means that the rate of change of the curve at Q is −269 volts per second.

This is the same as saying that the e.m.f. at $t = 0.1$ seconds is decreasing at the rate of 269 volts per second.

EXAMPLE 10.10

The formula $i = 2(1 - e^{-10t})$ gives the relationship between the instantaneous current i amperes and the time t seconds in an inductive circuit. Plot a graph of i against t taking values of t from 0 to 0.3 seconds at intervals of 0.05 seconds. Hence find:

a) the initial rate of growth of the current i when $t = 0$, and
b) the time taken for the current to increase from 1 to 1.6 amperes.

The curve is shown plotted in Fig. 10.23 from values obtained from a calculator.

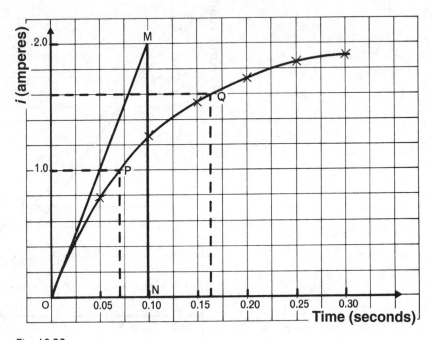

Fig. 10.23

a) When $t = 0$ the initial rate of growth will be given by the gradient of the tangent at O. The tangent at O is the line OM and its gradient may be found by using a suitable right-angled triangle at MNO and finding the ratio $\dfrac{MN}{ON}$.

Hence the initial rate of growth of $\quad i = \dfrac{MN}{ON} = \dfrac{2 \text{ amperes}}{0.1 \text{ seconds}}$

$= 20$ amperes per second

b) The point P on the curve corresponds to a current of 1.0 amperes and the time at which this occurs may be read from the t scale and is 0.07 seconds.

Similarly point Q corresponds to a 1.6 ampere current and occurs at 0.16 seconds.

Hence the time between P and Q $= 0.16 - 0.07 = 0.09$ seconds.

This means that the time for the current to increase from 1 to 1.6 amperes is 0.09 seconds.

Exercise 10.5

1) Plot a graph of $y = e^{2x}$ for values of x from -1 to $+1$ at 0.25 unit intervals. Use the graph to find the value of y when $x = 0.3$, and the value of x when $y = 5.4$.

2) Using values of x from -4 to $+4$ at one unit intervals plot a graph of $y = e^{-x/2}$. Hence find the value of x when $y = 2$, and the gradient of the curve when $x = 0$.

3) For a constant pressure process on a certain gas the formula connecting the absolute temperature T and the specific entropy s is $T = 24\,e^{3s}$. Plot a graph of T against s taking values of s equal to 1.000, 1.033, 1.066, 1.100, 1.133, 1.166 and 1.200. Use the graph to find the value of:

   a) T when $s = 1.09$ b) s when $T = 700$

4) The equation $i = 2.4e^{-6t}$ gives the relationship between the instantaneous current, i mA, and the time, t seconds. Plot a graph of i against t for values of t from 0 to 0.6 seconds at 0.1 second intervals. Use the curve obtained to find the rate at which the current is decreasing when $t = 0.2$ seconds.

5) In a capacitive circuit the voltage v and the time t seconds are connected by the relationship $v = 240(1 - e^{-5t})$. Draw the curve of v against t for values of $t = 0$ to $t = 0.7$ seconds at 0.1 second intervals. Hence find:

   a) the time when the voltage is 140 volts, and
   b) the initial rate of growth of the voltage when $t = 0$

Graphical solution of equations

In common with most algebraic functions, simultaneous equations can be displayed as graphs. This enables us to find solutions—generally indicated by the point(s) where the lines or curves intersect. Quadratic equations, which occur from analysis of engineering situations fairly often, have a distinctive shape to their curves—again solutions may be found graphically as an alternative to the algebraic approach.

Graphical solution of two simultaneous linear equations

Since the solutions we require have to satisfy both the given equations, they will be given by the values of the coordinates of the point where the graphs of the equations intersect.

EXAMPLE 10.11

The use of Kirchoff's laws in an electrical circuit network has resulted in two equations containing currents i_1 and i_2:

$$i_1 - 2i_2 = 2 \qquad\qquad [1]$$

$$3i_1 + i_2 = 20 \qquad\qquad [2]$$

Find the values of these currents using a simultaneous graphical solution.

Equation [1] may be written as:

$$i_1 = 2i_2 + 2$$

and equation [2] may be written as:

$$i_1 = -\frac{1}{3}i_2 + \frac{20}{3}$$

Here we have i_1 instead of the usual y, and i_2 replacing x.

i_2	−3	0	3
$i_1 = 2i_2 + 2$	−4	2	8
$i_1 = -\dfrac{1}{3}i_2 + \dfrac{20}{3}$	7.7	6.7	5.7

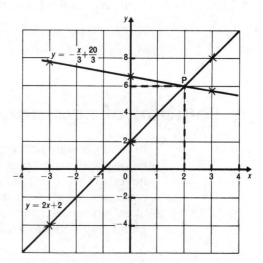

Fig. 10.24

The solutions of the equations will be given by the coordinates of the point where the two lines cross (that is, point P in Fig. 10.24). The coordinates of P are (2, 6).

Hence the two currents are $i_1 = 6$ and $i_2 = 2$.

Exercise 10.6

Solve the following equations:

1) $3x + 2y = 7$
   $x + y = 3$

2) $4x - 3y = 1$
   $x + 3y = 19$

3) The sum of the ages of two installations is 46 months. The modern version is 10 months younger than the original. Calculate their present ages.

4) A company's annual net profit of £8800 is divided amongst the two partners in the ratio $x : 2y$. If the first shareholder receives £2000 more than the other, find the values of x and y, and hence the respective shares of the profit.

5) The resistance R ohms of a wire at a temperature of $t\,°C$ is given by the formula $R = R_0(1 + \alpha t)$ where R_0 is the resistance at $0\,°C$ and α is a constant. The resistance is 35 ohms at a temperature of $80\,°C$ and 42.5 ohms at a temperature of $140\,°C$. Find R_0 and α. Hence find the resistance when the temperature is $50\,°C$.

6) A penalty clause states that a contractor will forfeit a certain sum of money for each day that he is late in completing a contract (i.e. the contractor gets paid the value of the original contract less any sum forfeit). If he is 6 days late he receives £5000 and if he is 14 days late he receives £3000. Find the amount of the daily forfeit and determine the value of the original contract.

7) The total cost of equipping two laboratories, A and B, is £30 000. If laboratory B costs £2000 more than laboratory A, find the cost of the equipment for each of them.

8) For a factory winch it is found that the effort E newtons and the load W newtons are connected by the equation $E = aW + b$. An effort of 90 N lifts a load of 100 N whilst an effort of 130 N lifts a load of 200 N. Find the values of a and b and hence determine the effort required to lift a load of 300 N.

Graphical solution of quadratic equations

An equation in which the highest power of the unknown is two, and containing no higher powers of the unknown, is called a quadratic equation. It is also known as an equation of the *second degree*. Thus

$$x^2 - 9 = 0 \qquad 2.5x^2 - 3.1x - 2 = 0$$
$$x^2 + 2x - 8 = 0 \qquad (x + 1)(x - 2) = 0$$

are all examples of quadratic equations.

Quadratic equations may be solved by plotting graphs: consider the graph of $y = ax^2 + bx + c$ which is in standard form. To solve the equation $ax^2 + bx + c = 0$ we have to find the values of x when $y = 0$, or the values of x where the graph cuts the x-axis.

EXAMPLE 10.12

The velocity, v, of the tool in a shaping machine, in terms of time, t, is given by the expression $v = t^2 - t - 2$. Find the times when the velocity is zero.

If we compare the equation here with that of the standard form, we have v in place of y, which will be plotted vertically, and t instead of x to be plotted horizontally.

No range of values has been suggested so we will, as usual, try $x = -4$ to $x = +4$:

t	-4	-3	-2	-1	0	1	2	3	4
t^2	16	9	4	1	0	1	4	9	16
$-t$	4	3	2	1	0	-1	-2	-3	-4
-2	-2	-2	-2	-2	-2	-2	-2	-2	-2
$v = t^2 - t - 2$	18	10	4	0	-2	-2	0	4	10

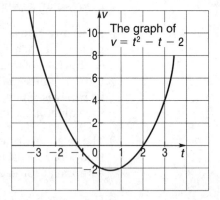

Fig. 10.25

Having marked out the axes, if we start plotting from the centre 'outwards' we shall soon realise that we can omit the extreme values (Fig. 10.25). The important portion of the curve is in the region of where it cuts the axes—especially the t-axis at $t = -1$ and $t = 2$, as for both these points $v = 0$.

Thus -1 and 2 are the times when velocity v is zero.

An alternative form of the quadratic equation

In Example 10.12 we found, in effect, that the quadratic equation $t^2 - t - 2 = 0$ had roots of -1 and 2.

If we examine the equation, we see that it factorises reasonably easily into: $(t + 1)(t - 2) = 0$ and we know from our earlier studies that the solutions of this equation are given by:

either $\qquad\qquad\qquad t + 1 = 0 \qquad$ or $\quad t - 2 = 0$

giving $\qquad\qquad\qquad\quad t = -1 \quad$ or $\quad t = 2$

These values confirm the results obtained graphically. So if you ever wish to study the graph of a quadratic equation in factorised form, do not multiply it out—it is better left as it is!

Exercise 10.7

By plotting suitable graphs solve the following equations:

1) $x^2 - 7x + 12 = 0$ (plot between $x = 0$ and $x = 6$)
2) $x^2 + 16 = 8x$ (plot between $x = 1$ and $x = 7$)
3) $x^2 - 9 = 0$ (plot between $x = -4$ and $x = 4$)
4) $x^2 + 2x - 15 = 0$
5) $3x^2 - 23x + 14 = 0$

6) The outer surface area of a small closed cylindrical metal container is given by 'twice the area of the base + the curved surface area' or $2\pi r^2 + 2\pi rh$. If the height of the container is 20 mm, and its surface area is 6000 mm^2, find its radius r in millimetres, by plotting a graph of a quadratic equation for r.

7) The phasor diagram representing two voltages v_1 and v_2 and their resultant voltage v_r is a triangle. Using the cosine rule gives:

$$v_r^2 = v_1^2 + v_2^2 - 2v_1v_2 \cos \theta$$

If $v_r = 1.45$ volts, $v_2 = 1.00$ volts and $\theta = 150°$ find a quadratic equation for v_1. We are only interested in the positive root, so plot a graph of y as a function of v_1 for values of v_1 from 0 to 1 volt, in 0.1 volt intervals. Hence find the voltage v_1.

Graphical solution of simultaneous linear and quadratic equations

Since the solutions we require have to satisfy both the given equations they will be given by the values of x and y where the graphs of the equations intersect.

EXAMPLE 10.13
Solve simultaneously the equations:

$$y = x^2 + 3x - 4$$
and
$$y = 2x + 4$$

We must first draw up tables of values, and will use the range $x = -4$ to $x = +4$:

x	-4	-2	0	2	4
x^2	16	4	0	4	16
$+3x$	-12	-6	0	6	12
-4	-4	-4	-4	-4	-4
$y = x^2 + 3x - 4$	0	-6	-4	6	24
$2x$	-8	-4	0	4	8
$+4$	4	4	4	4	4
$y = 2x + 4$	-4	0	4	8	12

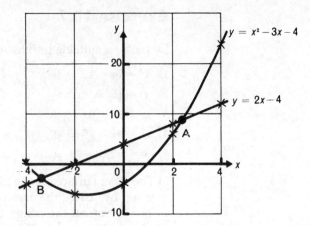

Fig. 10.26

The two graphs are shown plotted on the same axes in Fig. 10.26 and they intersect at the points A and B. Values of the x and y coordinates at these points will give the solutions of the given equations. We must be careful not to try and read the values too accurately, as the graphs have been plotted using only five values of x. In this case, even values to the first place of decimals cannot be guaranteed.

If more accurate answers are required then we must plot the portions of the graph containing points A and B using more values of x and also much larger scales.

Hence the required solutions are

x	2.4	-3.4
y	8.7	-2.7

Exercise 10.8

Solve simultaneously:

1) $y = x^2 - 2x - 2$
$x - y + 2 = 0$

2) $y = x^2 - x + 5$
$y = 2x + 5$

3) $y = 5x^2 + x - 3$
$y = 5x - 2$

4) $y = 2x^2 - 2.3x + 1$
$y = 3x - 0.25$

5) A rectangular template has a perimeter of 280 mm. The length of a diagonal drawn corner to corner is 100 mm. If x and y are the length and width of the template, draw a diagram and show that:

$$x + y = 140 \qquad \qquad [1]$$

and $$x^2 + y^2 = 10\,000 \qquad \qquad [2]$$

Hence find the dimensions of the template.

Trigonometric waveforms

The behaviour of alternating electrical currents and voltages is often aproached by looking at appropriate sine and cosine waveforms. Looking at 'pictures' of functions can often be more instructive than just pure number theory. Engineers often use their 'sixth sense' and get a 'feel' for what is happening from such displays—this is absent from cold algebraic manipulation.

Another use of waveforms is in work on vibration—maybe from the study of a motor vehicle's suspension or engine mounts.

Amplitude or peak value

The graphs of $\sin\theta$ and $\cos\theta$ each have a maximum value of $+1$ and a minimum value of -1.

Thus the graphs of $R\,\mathbf{sin}\,\theta$ and $R\,\mathbf{cos}\,\theta$ each have a maximum value of $+R$ and a minimum value of $-R$. These graphs are shown in Fig. 11.1.

The value of R is known as the **amplitude** or **peak value**.

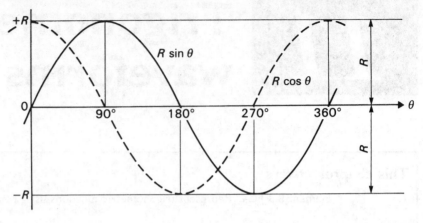

Fig. 11.1

Raising or lowering the axis of symmetry

We must first recognise $2 \sin \theta + 3$ as $(2 \sin \theta) + 3$. The effect of the constant $+3$ is to raise the waveform $2 \sin \theta$ 'bodily' (Fig. 11.2) so that its horizontal axis of symmetry is the $+3$ line.

Beware! $2 \sin \theta + 3$ is NOT $2 \sin(\theta + 3)$

This is a totally different expression as you will see later.

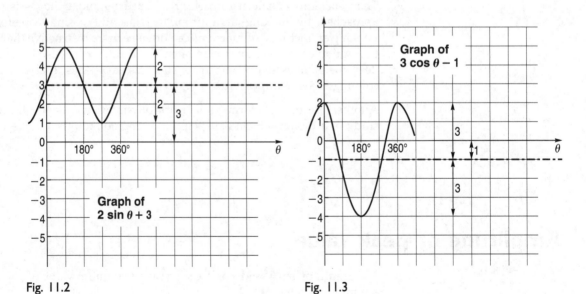

Fig. 11.2

Fig. 11.3

Similarly in Fig. 11.3 the waveform $3\cos \theta - 1$ is the curve of $3\cos \theta$ moved 'bodily' downwards so that its horizontal axis of symmetry is the -1 line.

Graphs of common sine and cosine functions

Graphs of $\sin\theta$, $\sin 2\theta$, $2\sin\theta$ and $\sin\frac{\theta}{2}$

Curves of the above trigonometrical functions are shown plotted in Fig. 11.4. You may find it useful to construct the curves using values obtained from a calculator.

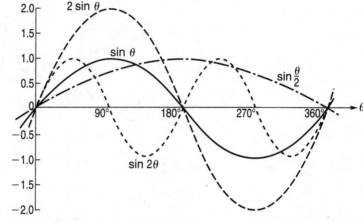

Fig. 11.4

Graphs of $\cos\theta$, $\cos 2\theta$, $2\cos\theta$ and $\cos\frac{\theta}{2}$

Cosine graphs are similar in shape to sine curves. You should plot graphs of the above functions from 0° to 360° using values obtained from a calculator.

Graphs of $\sin^2\theta$ and $\cos^2\theta$

It is sometimes necessary in engineering applications, such as when finding the root mean square value of alternating currents and voltages, to be familiar with the curves $\sin^2\theta$ and $\cos^2\theta$.

Values of the functions can be obtained using a calculator and their graphs are shown in Fig. 11.5. We should note that the curves are wholly positive, since squares of negative or positive values are always positive.

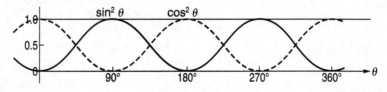

Fig. 11.5

Relation between angular and time scales

In Fig. 11.6 OP represents a radius, of length R, which rotates at a uniform angular velocity ω radians per second about O, the direction of rotation being anticlockwise.

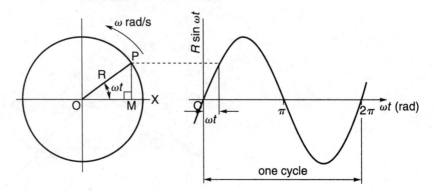

Fig. 11.6

Now

$$\text{Angular velocity} = \frac{\text{Angle turned through}}{\text{Time taken}}$$

∴ Angle turned through = Angular velocity × Time taken

and hence after a time t seconds

$$\theta = \omega t \qquad \text{radians}$$

Also from the rignt-angled triangle OPM:

$$\frac{\text{PM}}{\text{OP}} = \sin \text{POM}$$

∴ $$\text{PM} = \text{OP} \sin \text{POM}$$

or $$\text{PM} = R \sin \omega t$$

If a graph is drawn, as in Fig. 11.6, showing how PM varies with the angle ωt, the sine wave representing $R \sin \omega t$ is obtained. It can be seen that the peak value of this sine wave is R (i.e. the magnitude of the rotating radius).

The horizontal scale shows the angle turned through, ωt, and the waveform is said to be plotted on an **angular** or **ωt base**.

The waveforms represented by $R \cos \omega t$ are similar to sine waveforms, R being the peak value and $\dfrac{2\pi}{\omega}$ the period.

Cycle, period, frequency

Cycle

A cycle is the portion of the waveform which shows its complete shape without any repetition. It may be seen from Fig. 11.6 that one cycle is completed whilst the radius OP turns through 360° or 2π radians.

Period

The period is the time taken for the waveform to complete one cycle.

It will also be the time taken for OP to complete one revolution or 2π radians.

Now we know that

$$\text{Time taken} = \frac{\text{Angle turned through}}{\text{Angular velocity}}$$

Hence   The period $= \dfrac{2\pi}{\omega}$ seconds

Frequency

The number of cycles per second is called the frequency. The unit of frequency representing one cycle per second is the hertz (Hz).

Now if   1   cycle is completed in $\dfrac{2\pi}{\omega}$ seconds (a period)

*u*then $1 \div \dfrac{2\pi}{\omega}$ cycles are completed in 1 second

and therefore $\dfrac{\omega}{2\pi}$ cycles are completed in 1 second

Hence   Frequency $= \dfrac{\omega}{2\pi}$ Hz

Also since   Period $= \dfrac{2\pi}{\omega}$ s

Frequency $= \dfrac{1}{\text{Period}}$

Phasors and phase angle

The principal use of sine and cosine waveforms occurs in electrical technology in which they represent alternating currents and voltages.

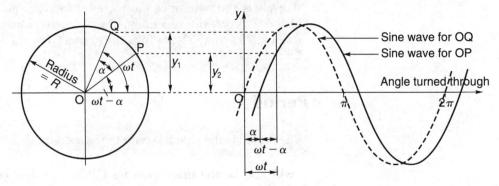

Fig. 11.7

In a diagram such as that shown in Fig. 11.7 the rotating radii OP and OQ are called phasors.

Fig. 11.7 shows two phasors OP and OQ, separated by an angle *a*, rotating at the same angular speed in an anticlockwise direction. The sine waves produced by OP and OQ are identical curves but they are displaced from each other. The amount of displacement is known as the phase difference and, measured along the horizontal axis, is *a*. The angle *a* is called the phase angle.

In Fig. 11.7 the phasor OP is said to lag behind phasor OQ by the angle *a*. If the radius of the phasor circle is R then OP = OQ = R and hence

for the phasor OQ, $\qquad\qquad\qquad y_1 = R \sin \omega t$
and for the phasor OP, $\qquad\qquad y_2 = R \sin (\omega t - a)$

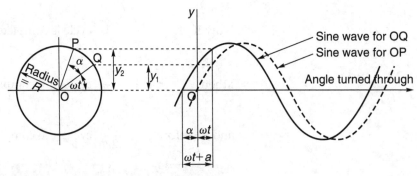

Fig. 11.8

Similarly in Fig. 11.8 the phasor OP leads the phasor OQ by the phase angle *a*.

Hence for the phasor OQ, $\qquad\qquad y_1 = R \sin \omega t$
and for the phasor OP, $\qquad\qquad y_2 = R \sin (\omega t + a)$

In practice it is usual to draw waveform on an 'angular' or '*ωt*' base when considering phase angles, as in the following example.

EXAMPLE 11.1
Sketch the waveforms of $\sin \omega t$ and $\sin \left(\omega t - \dfrac{\pi}{3} \right)$ on an angular base and identify the phase angle.

The curve $\sin \omega t$ will be plotted between $\omega t = 0$ and $\omega t = 2\pi$ radians (i.e. over 1 cycle).

Also $\sin \left(\omega t - \dfrac{\pi}{3} \right)$ will be plotted between values given by

$$\omega t - \frac{\pi}{3} = 0 \quad \text{and when} \quad \omega t - \frac{\pi}{3} = 2\pi$$

i.e. $\quad \omega t = \dfrac{\pi}{3}$ radians and $\quad \omega t = 2\pi + \dfrac{\pi}{3} = \dfrac{7\pi}{3}$ radians

The graphs are shown in Fig. 11.9.

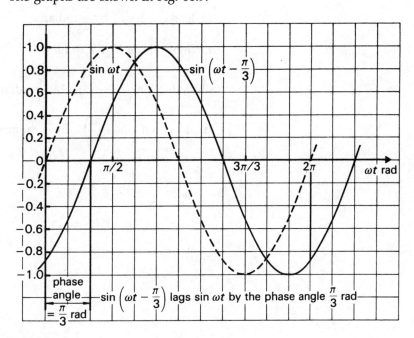

Fig. 11.9

Exercise 11.1

1) Plot the graphs of $\sin\theta$ and $\sin(\theta + 0.9)$ on the same axes on an angular base using radian units. Indicate the phase angle between the waveforms, and explain whether it is an angle of lead or lag.

2) Plot the curves of $\sin\theta$ and $\cos\theta$ on the same axes on an angular base showing a complete cycle of each waveform. Identify the phase angle between the curves.

3) Plot the graphs of $\sin\left(\omega t + \dfrac{\pi}{6}\right)$ and $\sin\omega t$ on the same axes on an angular base showing a cycle of each waveform. Indicate the phase angle between the curves, showing whether it is an angle of lead or lag.

4) Plot the graphs of $\sin\left(\omega t + \dfrac{\pi}{3}\right)$ and $\sin\left(\omega t + \dfrac{\pi}{4}\right)$ on the same axes, showing a complete cycle of each waveform, identifying the phase angle.

5) Write down the equation of the waveform which:

    a) leads $\sin\omega t$ by $\dfrac{\pi}{2}$ radians

    b) lags $\sin\omega t$ by π radians

    c) leads $\sin\left(\omega t - \dfrac{\pi}{3}\right)$ by $\dfrac{\pi}{3}$ radians

    d) lags $\sin\left(\omega t + \dfrac{\pi}{6}\right)$ by $\dfrac{\pi}{3}$ radians

Non-linear laws reducible to linear form

Why should we be so interested in reducing data to linear form that we devote a complete chapter to it? It is simply that we, as technologists, like to deal with straight line graphs. They are simpler to plot than the graphs of more complicated expressions and numerical results are obtained easily and quickly.

One section of this chapter concentrates on using the familiar Cartesian or rectangular coordinates. The other, although using rectangular axes, uses logarithmic scales in order to achieve a straight line plot. Here we have a good reason to fully appreciate how useful the 'tool' of logarithms can be!

Reduction to linear form

In this section we shall have practice in making direct substitutions for the function of x, and this enables straight line graphs to be drawn—an extremely straightforward method of reducing a difficult curve to something with which we are happier.

Equations of standard form $y = ax^n + b$

If we plot directly any equations of the form $y = ax^n + b$ we will obtain a curved graph.

However, if we plot x^n values along the horizontal axis instead of the usual x values, we will obtain a straight line. This assumes, of course, that the given data satisfies the original equation.

The form $\quad y = \dfrac{a}{x} + b$

Let $z = \dfrac{1}{x}$ so that the equation becomes $y = az + b$. If we now plot values of y against the corresponding values of z we will get a straight line since $y = az + b$ is of the standard linear form. In effect y has been plotted against $\dfrac{1}{x}$. The following example illustrates this method.

EXAMPLE 12.1

An experiment connected with the flow of water over a rectangular weir gave the following results:

C	0.503	0.454	0.438	0.430	0.425	0.421
H	0.1	0.2	0.3	0.4	0.5	0.6

The relation between C and H is thought to be of the form $C = \dfrac{a}{H} + b$. Test if this is so and find the values of the constants a and b.

In the suggested equation C is the sum of two terms, the first of which varies as $\dfrac{1}{H}$. If the equation $C = \dfrac{a}{H} + b$ is correct then when we plot C against $\dfrac{1}{H}$ we should obtain a straight line. To do this we draw up the following table:

C	0.503	0.454	0.438	0.430	0.425	0.421
$\dfrac{1}{H}$	10.00	5.00	3.33	2.50	2.00	1.67

The graph obtained is shown in Fig. 12.1. It is a straight line and hence the given values follow a law of the form $C = \dfrac{a}{H} + b$.

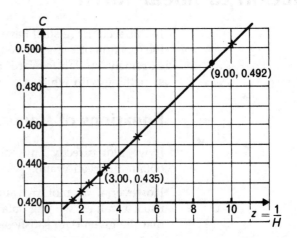

Fig. 12.1

To find the values of a and b we choose two points which *lie on the straight line*.

The point (3.00, 0.435) lies on the line.

$$\therefore \qquad 0.435 = 3.00a + b \qquad\qquad [1]$$

The point (9.00, 0.492) also lies on the line.

$$\therefore \qquad 0.492 = 9.00a + b \qquad\qquad [2]$$

Subtracting equation [1] from equation [2] gives

$$0.492 - 0.435 = a(9.00 - 3.00)$$
$$\therefore \qquad a = 0.0095$$

Substituting this value for a in equation [1] gives

$$0.435 = 3.00 \times 0.0095 + b$$
$$\therefore \qquad b = 0.435 - 0.0285 = 0.407$$

Hence the values of a and b are 0.0095 and 0.407, respectively.

The form $y = ax^2 + b$

Let $z = x^2$ and as previously if we plot values of y against z (in effect x^2) we will get a straight line since $y = az + b$ is of the standard form. The following example illustrates this method.

EXAMPLE 12.2

The fusing current I amperes for wires of various diameters d mm is as shown below:

d (mm)	5	10	15	20	25
I (amperes)	6.25	10	16.25	25	36.25

It is suggested that the law $I = ad^2 + b$ is true for the range of values given, a and b being constants. By plotting a suitable graph show that this law holds and from the graph find the constants a and b. Using the values of these constants in the equation $I = ad^2 + b$ find the diameter of the wire required for a fusing current of 12 amperes.

By putting $z = d^2$ the equation $I = ad^2 + b$ becomes $I = az + b$ which is the standard form of a straight line. Hence by plotting I against d^2 we should get a straight line if the law is true. To try this we draw up a table showing corresponding values of I and d^2.

d	5	10	15	20	25
$z = d^2$	25	100	225	400	625
I	6.25	10	16.25	25	36.25

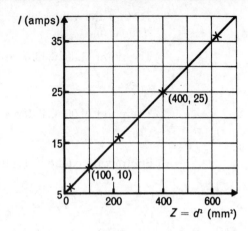

Fig. 12.2

From the graph (Fig. 12.2) we see that the points do lie on a straight line and hence the values obey a law of the form $I = ad^2 + b$.

To find the values of a and b choose two points which lie on the line and find their coordinates.

The point (400, 25) lies on the line

$$\therefore \qquad 25 = 400a + b \qquad\qquad [1]$$

The point (100, 10) lies on the line

$$\therefore \qquad 10 = 100a + b \qquad\qquad [2]$$

Subtracting equation [2] from equation [1] gives

$$15 = 300a$$
$$\therefore \qquad a = 0.05$$

Substituting $a = 0.05$ in equation [2] gives

$$10 = 100 \times 0.05 + b$$
$$\therefore \qquad b = 5$$

Therefore the law is $\quad I = 0.05d^2 + 5$

When $I = 12$ $\qquad 12 = 0.05d^2 + 5$

$$\therefore \qquad d = \sqrt{140} = 11.8 \text{ mm}$$

The form $\quad y = \dfrac{a}{x^2} + b$

Let $z = \dfrac{1}{x^2}$ so that the equation becomes $y = az + b$. If we now plot values of y against corresponding values of z we will get a straight line since $y = az + b$ is of the standard linear form. In effect y has been plotted against $\dfrac{1}{x^2}$.

The form $\quad y = a\sqrt{x} + b$

Let $z = \sqrt{x}$ and as previously, if we plot values of y against z (in effect $\sqrt{x}$) we will obtain a straight line since $y = az + b$ is of the standard linear form.

The form $y = \dfrac{a}{\sqrt{x}} + b$

This may also be written equivalently as $y = \dfrac{a}{x^{1/2}} + b$ or $y = ax^{-1/2} + b$.

Let $z = \dfrac{1}{\sqrt{x}}$ and, as previously, if we plot values of y against z $\left(\text{in effect } \dfrac{1}{\sqrt{x}}\right)$ we will obtain a straight line since $y = az + b$ is of the standard linear form.

Exercise 12.1

1) The following readings were taken during a test:

R (ohms)	85	73.3	64	58.8	55.8
I (amperes)	2	3	5	8	12

R and I are thought to be connected by an equation of the form $R = \dfrac{a}{I} + b$. Verify that this is so by plotting R (y-axis) against $\dfrac{1}{I}$ (x-axis) and hence find values for a and b.

2) In the theory of the moisture content of thermal insulation efficiency of porous materials the following table gives values of μ, the diffusion constant of the material, and k_m, the thermal conductivity of damp insulation material:

μ	1.3	2.7	3.8	5.4	7.2	10.0
k_m	0.0336	0.0245	0.0221	0.0203	0.0192	0.0183

Find the equation connecting μ and k_m if it is of the form $k_m = a + \dfrac{b}{\mu}$ where a and b are constants.

3) The accompanying table gives the corresponding values of the pressure, p, of mercury and the volume, v, of a given mass of gas at constant temperature.

p	90	100	130	150	170	190
v	16.66	13.64	11.54	9.95	8.82	7.89

By plotting p against the reciprocal of v obtain some relation between p and v. Evaluate any constants used in your method.

4) In an experiment, the resistance R of copper wire of various diameters d mm was measured and the following readings were obtained.

d (mm)	0.1	0.2	0.3	0.4	0.5
R (ohms)	20	5	2.2	1.3	0.8

Show that $R = \dfrac{k}{d^2}$ and find a suitable value for k.

5) The following table gives the thickness T mm of a brass flange brazed to a copper pipe of internal diameter D mm:

T (mm)	15.5	17.8	19.5	20.9	22.2	23.3
D (mm)	50	100	150	200	250	300

Show that T and D are connected by an equation of the form $T = a\sqrt{D} + b$, find the values of constants a and b, and find the thickness of the flange for a 70 mm diameter pipe.

6) The table shows how the coefficient of friction, μ, between a belt and a pulley varies with the speed, v m/s, of the belt. By plotting a graph show that $\mu = m\sqrt{v} + c$ and find the values of constants m and c.

μ	0.26	0.29	0.32	0.35	0.38
v	2.22	5.00	8.89	13.89	20.00

7) Using the table below show that the values are in agreement with the law $y = \dfrac{m}{\sqrt{x}} + c$. Hence evaluate the constants m and c.

x	0.2	0.8	1.2	1.8	2.5	4.4
y	1.62	1.51	1.49	1.47	1.46	1.44

Reduction to logarithmic form

Not all equations may be reduced to linear form by making a relatively straightforward substitution. Some types need special treatment and we aproach these by using our knowledge of logarithms and then plotting the values, given in the particular problem, on special graph paper having a logarithmic scale along one, or both, of the axes.

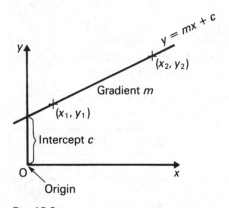

Fig. 12.3

The straight line

Remember the fundamental equation of the straight line:

$$y = mx + c$$

If m and c cannot be found directly from inspection of the graph, two points (x_1, y_1) and (x_2, y_2) on the line are chosen (Fig. 12.3) and their values substituted in the linear equation, giving the simultaneous equations:

$$y_1 = mx_1 + c$$
$$y_2 = mx_2 + c$$

from which m and c can be found.

Equations of the type $z = at^n$, $z = ab^t$ and $z = ae^{bt}$

In all the work which follows in this chapter the logarithms used will be to the base 10 and are denoted by 'lg'.

Consider the following relationships in which z and t are the variables, whilst a, b and n are constants.

$z = at^n$

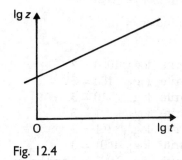

Fig. 12.4

Now $$z = at^n$$
and taking logs $$\lg z = \lg(at^n)$$
$$= \lg t^n + \lg a$$
$$\lg z = n \lg t + \lg a$$

The given values of the variables will satisfy this equation if they satisfy the original equation. Comparing this equation with $y = mx + c$, which is the standard equation of a straight line, we see that if we plot $\lg z$ on the y-axis and $\lg t$ on the x-axis the result will be a straight line (Fig. 12.4) and the values of the constants n and a may be found using the two point method described earlier.

$z = ab^t$

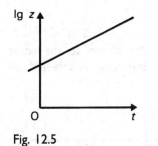

Fig. 12.5

Now $$z = ab^t$$
and taking logs $$\lg z = \lg(ab^t)$$
$$= \lg b^t + \lg a$$
$$\lg z = (\lg b)t + \lg a$$

We now proceed in a manner similar to that used for the previous equation by plotting $\lg z$ on the y-axis and t on the x-axis, and again obtain a straight line (Fig. 12.5).

$z = ae^{bt}$

Now $$z = ae^{bt}$$
and taking logs $$\lg z = \lg(ae^{bt})$$
$$= \lg e^{bt} + \lg a$$
$$= (b \lg e)t + \lg a$$
but $\lg e = 0.4343$

$\therefore$ $$\lg z = (0.4343b)t + \lg a$$

Again proceeding in a manner similar to that used for the previous equation, we plot $\lg z$ on the y-axis and t on the x-axis and again obtain a straight line (Fig. 12.5).

Logarithmic scales

In your previous work you may have used ordinary graph paper for problems involving logarithmic and exponent laws. This entailed finding the logs of each individual given number on at least one of the axes. However if logarithmic scales are used it is no longer necessary to find individual logs.

We use logs to the base 10 since, as you will see, natural logs to the base e would result in inconvenient numbers of the scales.

It has been shown earlier that : $\quad$ Number $=$ Base^{Logarithm}

and if we use a base of 10 then : $\quad$ Number $= 10^{\text{Logarithm}}$

Since $\quad\quad 1000 = 10^3 \quad$ then we may write $\log_{10} 1000 = 3$
and since $\quad 100 = 10^2 \quad$ then we may write $\log_{10} \quad 100 = 3$
and since $\quad\quad 10 = 10^1 \quad$ then we may write $\log_{10} \quad\quad 10 = 3$
and since $\quad\quad\quad 1 = 10^0 \quad$ then we may write $\log_{10} \quad\quad\quad 1 = 3$
and since $\quad 0.1 = 10^{-1} \quad$ then we may write $\log_{10} \quad 0.1 = 3$
and since $\quad 0.01 = 10^{-2} \quad$ then we may write $\log_{10} 0.01 = 3$
and since $0.001 = 10^{-3} \quad$ then we may write $\log_{10}0.001 = 3$ and so on.

These logarithms may be shown on a scale as shown in Fig. 12.6.

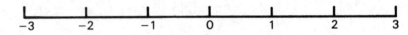

Fig. 12.6

However, since we wish to plot numbers directly on to the scale (without any reference to their logarithms) the scale is labelled as shown in Fig. 12.7.

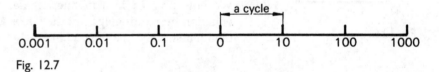

Fig. 12.7

Each division is called a cycle and is sub-divided using a logarithmic scale as, for instance, the scales on a slide rule. Two such cycles are shown in Fig. 12.8.

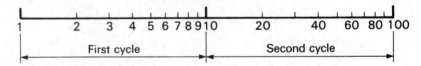

Fig. 12.8

The choice of numbers on the scale depends on the numbers allocated to the variable in the problem to be solved. Thus in Fig. 12.8 the numbers run from 1 to 100.

Logarithmic graph paper

Logarithmic scales may be used on graph paper in place of the more usual linear scales. By using graph paper ruled in this way log plots may be made without the necessity of looking up the logs of each given value. Semi-logarithmic graph paper is also available and has one way ruled with log scales whilst the other way has the usual linear scale. Examples of each are shown in Figs 12.9 and 12.10.

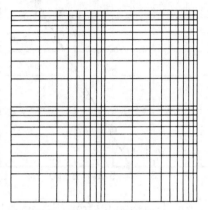

Full logarithmic rulings

Fig. 12.9

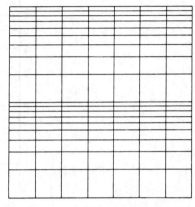

Semi-logarithmic rulings

Fig. 12.10

The use of these scales and the special graph paper is shown by the examples which follow.

EXAMPLE 12.3

The law connecting two quantities z and t is of the form $z = at^n$. Find the law given the following pairs of values:

z	3.170	4.603	7.499	10.50	15.17
t	7.980	9.863	13.03	15.81	19.50

The relationship gives (see text p. 239)

$$z = at^n$$
$$\lg z = n \lg t + \lg a \qquad [1]$$

Hence we plot the given values of z and t on log scales as shown in Fig. 12.11.

On both the vertical and horizontal axes we require 2 cycles, the first for values from 1 to 10 and the second for values from 10 to 100.

The constants are found by taking two pairs of cordinates:

Point (25, 23) lies on the line, and putting these values in Equation [1],

$$\lg 23 = n \lg 25 + \lg a \qquad [2]$$

Point (4.1, 1) lies on the line and putting these values in Equation [1],

$$\lg 1 = n \lg 4.1 + \lg a \qquad [3]$$

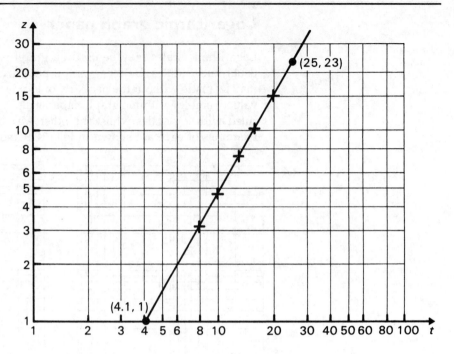

Fig. 12.11

Subtracting Equation [3] from Equation [2],

$$\lg 23 - \lg 1 = n(\lg 25 - \lg 4.1)$$
$$\lg (23/1) = n[\lg (25/4.1)]$$
$$n = \frac{\lg (23/1)}{\lg (25/4.1)}$$
$$n = 1.73$$

Substituting this value of n in Equation [3],

$$\lg 1 = 1.73 \lg 4.1 + \lg a$$
$$\therefore \qquad \lg a = \lg 1 - 1.73 \lg 4.1$$
$$\therefore \qquad a = 0.087$$

Hence the law is $\quad z = 0.87t^{1.73}$

EXAMPLE 12.4
The table gives values obtained in an experiment. It is thought that the law may be of the form $z = ab^t$, where a and b are constants. Verify this and find the law.

t	0.190	0.250	0.300	0.400
z	11 220	18 620	26 920	61 660

We think that the relationship is of the form:

$$z = ab^t$$

which gives (see text p. 239),

$$\lg z = (\lg b)t + \lg a \qquad [1]$$

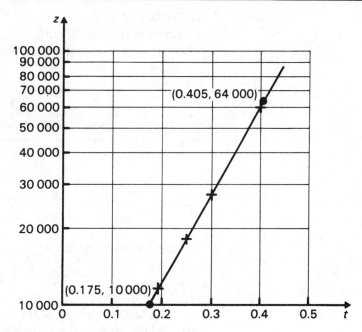

Fig. 12.12

Hence we plot the given values of z on a vertical log scale—the t values, however, will be on the horizontal axis on an ordinary linear scale (Fig. 12.12).

The points lie on a straight line, and hence the given values of z and t obey the law. We now have to find the cordinates of two points lying on the line.

Point (0.405, 64 000) lies on the line, and substituting in Equation [1],

$$\lg 64\,000 = (\lg b)0.405 + \lg a \qquad [2]$$

Point (0.175, 10 000) lies on the line, and substituting in Equation [1],

$$\lg 10\,000 = (\lg b)0.175 + \lg a \qquad [3]$$

Subtracting Equation [3] from Equation [2],

$$\lg 64\,000 - \lg 10\,000 = (\lg b)(0.405 - 0.175)$$
$$\therefore \qquad 4.8062 - 4.0000 = 0.230(\lg b)$$
$$\therefore \qquad \lg b = \frac{0.8062}{0.230} = 3.5052$$
$$\therefore \qquad b = 3200$$

Sustituting in the Equation [3],

$$\lg 10\,000 = (3.5052)0.175 + \lg a$$
$$\therefore \qquad \lg a = \lg 10\,000 - 3.5052(0.175)$$
$$= 4 - 0.6134 = 3.3866$$
$$\therefore \qquad a = 2436$$

Hence the law is:

$$z = 2436(3200)^t$$

EXAMPLE 12.5

V and t are connected by the law $V = ae^{bt}$. If the values given in the table satisfy the law, find the constants a and b.

t	0.05	0.95	2.05	2.95
v	20.70	24.49	30.27	36.06

The law is:

$$V = ae^{bt}$$

which gives (see text p. 239),

$$\lg V = 0.4343bt + \lg a \qquad [1]$$

As in the last example V values are plotted on a log scale on the vertical axis, whilst the t values are plotted on the horizontal axis on an ordinary linear scale (Fig. 12.13).

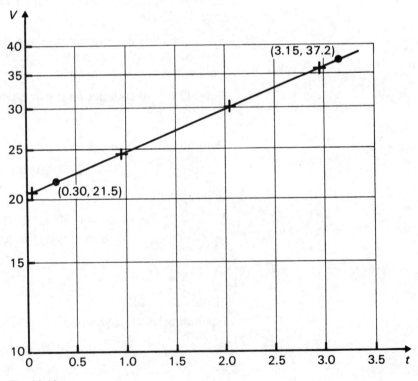

Fig. 12.13

Point (3.15, 37.2) lies on the line, and substituting in Equation [1],

$$\lg 37.2 = (0.4343)b(3.15) + \lg a \qquad [2]$$

Point (0.30, 21.5) lies on the line, and substituting in Equation [1],

$$\lg 21.5 = (0.4343)b(0.30) + \lg a \qquad [3]$$

Subtracting, Equation [3] from Equation [2],

$$\lg 37.2 - \lg 21.5 = (0.4343)b(3.15 - 0.30)$$

$$\therefore \qquad b = \frac{\lg (37.2/21.5)}{(0.4343)(2.85)}$$

Thus $\qquad b = 0.192$

Substituting in Equation [3]

$$\lg 21.5 = (0.4343)(0.192)(0.30) + \lg a$$

$$\therefore \qquad \lg a = 1.3324 - 0.0250$$

$$\therefore \qquad a = 20.3$$

Exercise 12.2

1) Using log–log graph paper show that the following set of values for x and y follows a law of the type $y = ax^n$. From the graph determine the values of a and n.

x	4	16	25	64	144	296
y	6	12	15	24	36	52

2) The following results were obtained in an experiment to find the relationship between the luminosity I of a metal filament lamp and the voltage V:

V	60	80	100	120	140
I	11	20	89	186	319

Allowing for the fact that an error was made in one of the readings show that the law between I and V is of the form $I = aV^n$ and find the probable correct value of the reading. Find the value of n.

3) Two quantities t and m are plotted on log–log graph paper, t being plotted vertically and m being plotted horizontally. The result is a straight line and from the graph it is found that:

when $\qquad m = 8, \qquad t = 6.8$

and when $\qquad m = 20, \qquad t = 26.9$

Find the law connecting t and m.

4) The intensity of radiation R from certain radioactive materials at a particular time t is thought to follow the law $R = kt^n$. In an experiment to test this the following values were obtained:

R	58	43.5	26.5	14.5	10
t	1.5	2	3	5	7

Show that the assumption was correct and evaluate k and n.

5) The values given in the following table are thought to obey a law of the type $y = ab^{-x}$. Check this statement and find the values of the constants a and b.

x	0.1	0.2	0.4	0.6	1.0	1.5	2.0
y	175	158	60	32	6.4	1.28	0.213

6) The force F on the tight side of a driving belt was measured for different values of the angle of lap θ and the following results were obtained:

F	7.4	11.0	17.5	24.0	36.0
θ rad	$\pi/4$	$\pi/2$	$3\pi/4$	π	$5\pi/4$

Construct a graph to show that these values conform approximately to an equation of the form $F = ke^{\mu\theta}$.
Hence find the constants μ and k.

7) A capacitor and resistor are connected in series. The current i amperes after time t seconds is thought to be given by the equation $i = Ie^{-t/T}$ where I amperes is the initial charging current and T seconds is the time constant. Using the following values verify the relationship and find the values of the constants I and T:

i amperes	0.0156	0.0121	0.00945	0.00736	0.00573
t seconds	0.05	0.10	0.15	0.20	0.25

8) For a constant pressure process on a certain gas the formula connecting the absolute temperature T and the specific entropy s is of the form $T = ke^{cs}$ where e is the logarithmic base and k and c are constants. When $T = 460$, $s = 1.000$, and when $T = 600$, $s = 1.089$. Find constants k and c to three significant figures.

9) The instantaneous e.m.f. v induced in a coil after a time t is given by $v = Ve^{-t/T}$, where V and T are constants. Find the values of V and T given the following values:

v	95	80	65	40	25
t	0.00013	0.00056	0.00108	0.00229	0.00347

13 Rates of change and differentiation

It is unfortunate that the word 'differentiation' has to be used before you have any inkling of what it is all about. However, being resolute (as you must be to have reached this part of the book!), we think you will have a pleasant surprise—most of us are happy when drawing a simple graph and taking measurements from it. Well this is exactly how you will begin and you will soon get the 'feel' of what it is all about. Notice how important the word 'feel' is to us as engineers. As this is your first meeting with this vast topic of calculus, and the time that can be spent is limited, you will not appreciate at this stage how useful it can be. However, this serves as a useful introduction and, with the extra knowledge from the final chapter, you will be well prepared when meeting calculus in your future studies.

Differentiation

Here we start by drawing a graph and finding a gradient by measurement. We then move to the idea of small quantities, and what happens to them as they become very tiny. This gives us another way of finding a gradient and we are soon on the way to using differentiation as another 'tool' in problem solving.

The gradient of a curve

Graphical method

In mathematics and technology we often need to know the rate of change of one variable with respect to another. For instance, velocity is the rate of change of distance with respect to time, and acceleration is the rate of change of velocity with respect to time.

Consider the graph of $y = x^2$, part of which is shown in Fig. 13.1. As the values of x increase so do the values of y, but they do not increase at the same rate. A glance at the portion of the curve shown shows that the values of y increase faster when x is large, because the gradient of the curve is increasing.

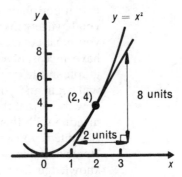

Fig. 13.1

To find the rate of change of y with respect to x at a particular point we need to find the gradient of the curve at that point.

If we draw a tangent to the curve at the point, the gradient of the tangent will be the same as the gradient of the curve.

EXAMPLE 13.1

Find the gradient of the curve $y = x^2$ at the point where $x = 2$

The point where $x = 2$ is the point (2, 4). We draw a tangent at this point, as shown in Fig. 13.1. Then by constructing a right-angled triangle the gradient is found to be $\frac{8}{2} = 4$. This gradient is positive, in accordance with our previous work, since the tangent slopes upwards from left to right.

EXAMPLE 13.2

Draw the graph of $y = x^2 - 3x + 7$ between $x = -4$ and $x = 3$ and hence find the gradient at: a) the point $x = -3$, b) the point $x = 2$.

a) At the point where $x = -3$,
$$y = (-3)^2 - 3(-3) + 7 = 25$$

At the point $(-3, 25)$, draw a tangent as shown in Fig. 13.2. The gradient is found by drawing a right-angled triangle (which should be as large as conveniently possible for accuracy) as shown, and measuring its height and base.

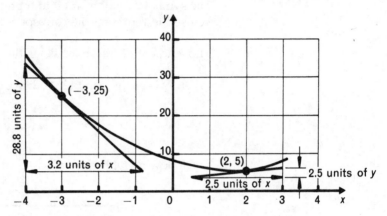

Fig. 13.2

Hence: $\qquad$ Gradient at point $(-3, 25) = -\dfrac{28.8}{3.2} = -9$

The negative sign indicating a downward slope from left to right.

b) At the point where $x = 2$,
$$y = 2^2 - 3(2) + 7 = 5$$

Hence by drawing a tangent and a right-angled triangle at the point $(2, 5)$ in a similar manner to above,

$$\text{Gradient at point } (2, 5) = \frac{2.5}{2.5} = 1$$

being positive as the tangent slopes upwards from left to right.

Exercise 13.1

1) Draw the graph of $2x^2 - 5$ for values of x between -2 and $+3$. Draw, as accurately as possible, the tangents to the curve at the points where $x = -1$ and $x = +2$ and hence find the gradient of the curve at these points.

2) Draw the curve $y = x^2 - 3x + 2$ from $x = 2.5$ to $x = 3.5$ and find its gradient at the point where $x = 3$.

3) Draw the curve $y = x - \dfrac{1}{x}$ from $x = 0.8$ to 1.2. Find its gradient at $x = 1$.

Numerical method

The gradient of a curve may always be found by graphical means but this method is often inconvenient. A numerical method will now be developed.

Consider the curve $y = x^2$, part of which is shown in Fig. 13.3. Let P be the point on the curve at which $x = 1$ and $y = 1$. Q is a variable point on the curve, which will be considered to start at the point (2, 4) and move down the curve towards P, rather like a bead slices down a wire.

The symbol δx will be used to represent an increment of x, and δy will be used to represent the corresponding increment of y.

The gradient of the chord PQ is then $\dfrac{\delta y}{\delta x}$

When Q is at the point (2, 4) then $\delta x = 1$, and $\delta y = 3$

$$\therefore \qquad \frac{\delta y}{\delta x} = \frac{3}{1} = 3$$

The following table shows how $\dfrac{\delta y}{\delta x}$ alters as Q moves nearer and nearer to P.

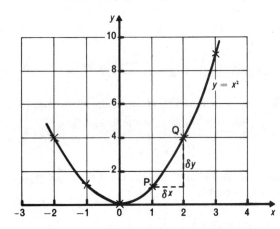

Fig. 13.3

Co-ordinates of Q				Gradient of
x	y	δy	δx	$PQ = \dfrac{\delta y}{\delta x}$
2	4	3	1	3
1.5	2.25	1.25	0.5	2.5
1.4	1.96	0.96	0.4	2.4
1.3	1.69	0.69	0.3	2.3
1.2	1.44	0.44	0.2	2.2
1.1	1.21	0.21	0.1	2.1
1.01	1.0201	0.0201	0.01	2.01
1.001	1.002001	0.002001	0.001	2.001

It will be seen that as Q approaches nearer and nearer to P, the value of $\dfrac{\delta y}{\delta x}$ approaches 2. It is reasonable to suppose that eventually when Q coincides with P (that is, when the chord PQ becomes a tangent to the curve at P) the gradient of the tangent will be exactly equal to 2. The gradient of the tangent will give us the gradient of the curve at P.

Now as Q approaches P, δx tends to zero and the gradient of the chord, $\dfrac{\delta y}{\delta x}$, tends, in the limit (as we say), to the gradient of the tangent. We denote the gradient of the tangent as $\dfrac{dy}{dx}$. We can write all this as

$$\underset{\delta x \to 0}{\text{Limit}} \; \frac{\delta y}{\delta x} = \frac{dy}{dx}$$

Differentiation from first principles

Instead of selecting special values for δy and δx let us now consider the general case, so that P has the coordinates (x, y) and Q has the coordinates $(x + \delta x, y + \delta y)$, (Fig. 13.4). Q is taken very close to P, so that δx is a very small quantity.

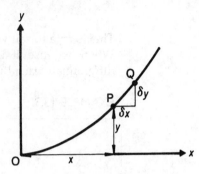

Fig. 13.4

Now $$y = x^2$$

and as $Q(x + \delta x, y + \delta y)$ lies on the curve, then

$$y + \delta y = (x + \delta x)^2$$
$$\therefore \qquad y + \delta y = x^2 + 2x\delta x + (\delta x)^2$$

But $y = x^2$, so $\qquad \delta y = 2x\delta x + (\delta x)^2$

and, by dividing both sides by δx, the gradient of chord PQ is

$$\frac{\delta y}{\delta x} = 2x + \delta x$$

As Q approaches P, δx tends to zero and $\dfrac{\delta y}{\delta x}$ tends, in the limit, to the gradient of the tangent of the curve at P.

Thus $$\underset{\delta x \to 0}{\text{Limit}} \frac{\delta y}{\delta x} = \frac{\mathrm{d}y}{\mathrm{d}x} = 2x$$

The process of finding $\dfrac{\mathrm{d}y}{\mathrm{d}x}$ is called *differentiation*.

The symbol $\dfrac{\mathrm{d}y}{\mathrm{d}x}$ means the differential coefficient of y with respect to x.

We can now check our assumption regarding the gradient of the curve at P. Since at P the value of $x = 1$, then substituting in the

expression $\qquad \dfrac{\mathrm{d}y}{\mathrm{d}x} = 2x \quad$ we get $\quad \dfrac{\mathrm{d}y}{\mathrm{d}x} = 2 \times 1 = 2$

and we see that our assumption was correct.

251

Differential coefficient of x^n

It can be shown, by a method similar to that used for finding the differential coefficient of x^2, that

$$\text{If} \qquad y = x^n \qquad \text{then} \qquad \frac{dy}{dx} = nx^{n-1}$$

This is true for all value of n including negative and fractional indices. When we use it as a formula it enables us to avoid having to differentiate each time from first principles.

EXAMPLE 13.3

$y = x^3$ $\therefore \dfrac{dy}{dx} = 3x^2$	$y = \dfrac{1}{x} = x^{-1}$ $\therefore \dfrac{dy}{dx} = -x^{-2} = -\dfrac{1}{x^2}$
$y = \sqrt{x} = x^{1/2}$ $\therefore \dfrac{dy}{dx} = \dfrac{1}{2}x^{-1/2} = \dfrac{1}{2}\dfrac{1}{x^{1/2}} = \dfrac{1}{2\sqrt{x}}$	$y = \sqrt[5]{x^2} = x^{2/5}$ $\therefore \dfrac{dy}{dx} = \dfrac{2}{5}x^{2/5-1} = \dfrac{2}{5}x^{-3/5} = \dfrac{2}{5(\sqrt[5]{x^3})}$

When a power of x is multiplied by a constant, that constant remains unchanged by the process of differentiation

$$\text{Hence if} \qquad y = ax^n \qquad \text{then} \qquad \frac{dy}{dx} = anx^{n-1}$$

EXAMPLE 13.4

$y = 2x^{1.3}$ $\therefore \dfrac{dy}{dx} = 2(1.3)x^{0.3} = 2.6x^{0.3}$	$y = \dfrac{1}{5}x^7$ $\therefore \dfrac{dy}{dx} = \dfrac{1}{5} \times 7x^6 = \dfrac{7}{5}x^6$
$y = \dfrac{3}{4}\sqrt[3]{x} = \dfrac{3}{4}x^{1/3}$ $\therefore \dfrac{dy}{dx} = \dfrac{3}{4} \times \dfrac{1}{3}x^{-2/3} = \dfrac{1}{4}x^{-2/3}$	$y = \dfrac{4}{x^2} = 4x^{-2}$ $\therefore \dfrac{dy}{dx} = 4(-2)x^{-3} = -8x^{-3}$

When a numerical constant is differentiated the result is zero

This can be seen since $x^0 = 1$ and we can write, for example, constant 4 as $4x^0$, then differentiating with respect to x we get $4(0)x^{-1} = 0$

If, as an alternative method, we plot the graph of $y = 4$ we get a straight line parallel with the x-axis as shown in Fig. 13.5.

The gradient of the line is zero: that is, $\dfrac{dy}{dx} = 0$

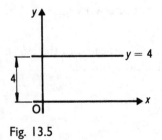

Fig. 13.5

To differentiate an expression containing the sum of several terms, differentiate each individual term separately

EXAMPLE 13.5

a)
$$y = 3x^2 + 2x + 3$$

$$\therefore \quad \frac{dy}{dx} = 3(2)x + 2(1)x^0 + 0 = 6x + 2$$

b)
$$y = ax^3 + bx^2 + cx + d \qquad \text{where } a, b, c \text{ and } d \text{ are constants,}$$

$$\therefore \quad \frac{dy}{dx} = 3ax^2 + 2bx + c$$

So far our differentiation has been in terms of x and y only. But they are only letters representing variables and we may choose other letters of symbols.

c)
$$s = \sqrt{t} + \frac{1}{\sqrt{t}} = t^{1/2} + t^{-1/2}$$

$$\therefore \quad \frac{ds}{dt} = \frac{1}{2}t^{-1/2} + \left(-\frac{1}{2}\right)t^{-3/2} = \frac{1}{2\sqrt{t}} - \frac{1}{2\sqrt{t^3}}$$

d)
$$v = 3.1u^{1.4} - \frac{3}{u} + 5 = 3.1u^{1.4} - 3u^{-1} + 5$$

$$\therefore \quad \frac{dv}{du} = (3.1)(1.4)u^{0.4} - 3(-1)u^{-2} = 4.34u^{0.4} + \frac{3}{u^2}$$

Finding the gradient of a curve by differentiation

EXAMPLE 13.6

Find the gradient of the graph $y = 3x^2 - 3x + 4$:

a) when $x = 3$, and, b) when $x = -2$

The gradient at a point is expressed by $\dfrac{dy}{dx}$

If
$$y = 3x^2 - 3x + 4$$

then
$$\frac{dy}{dx} = 6x - 3$$

a) When $x = 3$

$$\frac{dy}{dx} = 6(3) - 3 = 15$$

b) When $x = -2$

$$\frac{dy}{dx} = 6(-2) - 3 = -15$$

You may find this summary of the laws of indices useful when working through Exercise 13.2.

$$a^m \times a^n = a^{m+n}$$
$$\frac{a^m}{a^n} = a^{m-n}$$
$$(a^m)^n = a^{m \times n}$$
$$a^0 = 1$$
$$a^{-n} = \frac{1}{a^n}$$
$$\sqrt[n]{a} = a^{1/n}$$

Exercise 13.2

Differentiate the following:

1) $y = x^2$ 2) $y = x^7$ 3) $y = 4x^3$

4) $y = 6x^5$ 5) $s = 0.5t^3$ 6) $A = \pi R^2$

7) $y = x^{1/2}$ 8) $y = 4x^{3/2}$ 9) $y = 2 \times \sqrt{x}$

10) $y = 3 \times \sqrt[3]{x^2}$ 11) $y = \dfrac{1}{x^2}$ 12) $y = \dfrac{1}{x}$

13) $y = \dfrac{3}{5x}$ 14) $y = \dfrac{2}{x^3}$ 15) $y = \dfrac{1}{\sqrt{x}}$

16) $y = \dfrac{2}{3 \times \sqrt{x}}$ 17) $y = \dfrac{5}{x \times \sqrt{x}}$ 18) $s = \dfrac{3 \times \sqrt{t}}{5}$

19) $K = \dfrac{0.01}{h}$ 20) $y = \dfrac{5}{x^7}$

21) $y = 4x^2 - 3x + 2$ 22) $s = 3t^3 - 2t^2 + 5t - 3$

23) $q = 2u^2 - u + 7$ 24) $y = 5x^4 - 7x^3 + 3x^2 - 2x + 5$

25) $s = 7t^5 - 3t^2 + 7$ 26) $y = \dfrac{x + x^3}{\sqrt{x}}$

27) $y = \dfrac{3 + x^2}{x}$ 28) $y = \sqrt{x} + \dfrac{1}{\sqrt{x}}$ 29) $y = x^3 + \dfrac{3}{\sqrt{x}}$

30) $s = t^{1.3} - \dfrac{1}{4t^{2.3}}$ 31) $y = \dfrac{3x^3}{5} - \dfrac{2x^2}{7} + \sqrt{x}$

32) $y = 0.08 + \dfrac{0.01}{x}$ 33) $y = 3.1x^{1.5} - 2.4x^{0.6}$

34) $y = \dfrac{x^3}{2} - \dfrac{5}{x} + 3$ 35) $s = 10 - 6t + 7t^2 - 2t^3$

36) Find the gradient of the curve $y = 3x^2 + 7x + 3$ at the points where $x = -2$ and $x = 2$

37) Find the gradient of the curve $y = 2x^3 - 7x^2 + 5x - 3$ at the points where $x = -1.5$, $x = 0$ and $x = 3$

38) Find the values of x for which the gradient of the curve $y = 3 + 4x - x^2$ is equal to:

a) -1 b) 0 c) 2

Differentiation of trigonometric functions

It can be shown that, for the standard forms of trigonometric expressions, where ω and α are constants:

$$\frac{d}{dt}\sin(\omega t + \alpha) = \omega \cos(\omega t + \alpha)$$

$$\frac{d}{dt}\cos(\omega t + \alpha) = -\omega \sin(\omega t + \alpha)$$

EXAMPLE 13.7

Find $\dfrac{dy}{dt}$ if $y = \cos\left(2t + \dfrac{\pi}{2}\right)$

If we compare this with the standard form, then $\omega = 2$ and $\alpha = \dfrac{\pi}{2}$

Thus if $\qquad\qquad y = \cos\left(2t + \dfrac{\pi}{2}\right)$

then $\qquad\qquad \dfrac{dy}{dt} = -2\sin\left(2t + \dfrac{\pi}{2}\right)$

Differentiation of logarithmic functions

It can be shown that:

$$\frac{d}{dx}(\log_e x) = \frac{1}{x}$$

EXAMPLE 13.8

Find $\qquad\dfrac{d}{dx}(\log_e 2x)$

Let $\qquad\qquad y = \log_e 2x$

$\qquad\qquad\qquad = \log_e 2 + \log_e x \quad$ (using law 1 of logs
$\qquad\qquad\qquad\qquad\qquad\qquad\qquad\qquad$ for numbers multiplied)

Now $\log_e 2$ is simply a constant, so it will disappear on differentiation.

Thus if $\qquad y = \log_e 2 + \log_e x$

then $\qquad \dfrac{dy}{dx} = \dfrac{1}{x}$

Differentiation of exponent functions

It can be shown, where a is a constant, that:

$$\frac{\mathrm{d}}{\mathrm{d}x}(\mathrm{e}^{ax}) = a\mathrm{e}^{ax}$$

EXAMPLE 13.9

Find
$$\frac{\mathrm{d}}{\mathrm{d}t}(\mathrm{e}^{5t} + \mathrm{e}^{-2t})$$

As usual, when dealing with terms added together, we deal with each term individually.

Thus if we let
$$y = \mathrm{e}^{5t} + \mathrm{e}^{-2t}$$

then
$$\frac{\mathrm{d}y}{\mathrm{d}x} = 5\mathrm{e}^{5t} + (-2)(\mathrm{e}^{-2t})$$
$$= 5\mathrm{e}^{5t} - 2\mathrm{e}^{-2t}$$

EXAMPLE 13.10

Find
$$\frac{\mathrm{d}}{\mathrm{d}x}(\mathrm{e}^{x})$$

Here we have the simple exponent function e^x where constant $a = 1$.

Thus
$$\frac{\mathrm{d}}{\mathrm{d}x}(\mathrm{e}^{x}) = \mathrm{e}^{x}$$

This verifies the important property of the exponent function which we first discovered on page 215, namely:

The exponent function e^x has a differential coefficient of e^x, which is, therefore, equal to itself.

We may now summarise differential coefficients of the more common functions:

y	$\dfrac{\mathrm{d}}{\mathrm{d}x}$
ax^n	anx^{n-1}
$\sin(\omega t + \alpha)$	$\omega\cos(\omega t + \alpha)$
$\cos(\omega t + \alpha)$	$-\omega\sin(\omega t + \alpha)$
$\log_{\mathrm{e}}x$	$\dfrac{1}{x}$
e^{ax}	$a\mathrm{e}^{ax}$

Exercise 13.3

1) Find $\dfrac{\mathrm{d}}{\mathrm{d}\theta}(\sin 2\theta + \cos 5\theta)$

2) Find $\dfrac{\mathrm{d}}{\mathrm{d}t}\sin(4t + \pi)$

3) Find $\dfrac{\mathrm{d}}{\mathrm{d}t}\cos\left(7t - \dfrac{3\pi}{2}\right)$

4) Find $\dfrac{\mathrm{d}y}{\mathrm{d}x}$ if $y = \log_e 9x$ with the aid of law 1 of logs.

5) Find $\dfrac{\mathrm{d}y}{\mathrm{d}x}$ if $y = \log_e\left(\dfrac{10}{x}\right)$ with the aid of law 2 of logs.

6) Find $\dfrac{\mathrm{d}y}{\mathrm{d}x}$ if $y = \log_e(x^2)$ with the aid of law 3 of logs.

7) Find $\dfrac{\mathrm{d}}{\mathrm{d}u}(e^{3u} - e^{-3u})$

8) Find $\dfrac{\mathrm{d}}{\mathrm{d}v}(6e^{-5v})$

Application of calculus to problems

Here are three distinctly different applications of differentiation.

Distance, time, velocity and acceleration, either linear or angular, crop up all the time in engineering, and are well catered for.

What is the maximum power an engine can produce or what is the least cost of producing a certain component? Such questions occur frequently and can often be solved by the application of differentiation.

Distance, time, velocity and acceleration for linear and angular motion

Suppose that a vehicle starts from rest and travels 60 metres in 12 seconds. The average velocity may be found by dividing the total distance travelled by the total time taken, that is $\frac{60}{12} = 5\,\text{m/s}$. This is *not* the *instantaneous* velocity, however, *at* a time of 12 seconds, but is the *average velocity* over the 12 seconds as calculated previously.

Fig. 13.6 shows a graph of distance s against time t. The average velocity over a period is given by the gradient of the chord which meets the curve at the extremes of the period. Thus in the diagram the gradient of the dotted chord QR gives the average velocity between $t = 2\,\text{s}$ and $t = 6\,\text{s}$. It is found to be $\frac{13}{4} = 3.25\,\text{m/s}$.

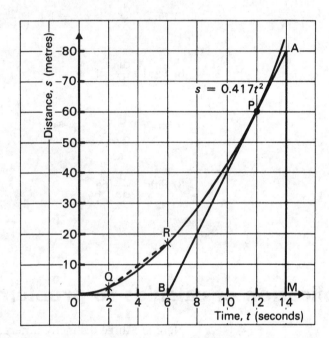

Fig. 13.6

The velocity at any point is the rate of change of s with respect to t and may be found by finding the gradient of the curve at that point. In mathematical notation that is given by $\dfrac{\mathrm{d}s}{\mathrm{d}t}$.

Suppose we know that the relationship between s and t is

$$s = 0.417t^2$$

Then velocity $\qquad v = \dfrac{\mathrm{d}s}{\mathrm{d}t} = 0.834t$

and hence when $t = 12$ seconds, $\quad v = 0.834 \times 12 = 10\,\text{m/s}$.

This result may be found graphically, Fig. 13.6, by drawing the tangent to the curve of s against t at the point P and constructing a suitable right-angled triangle ABM.

Hence the velocity at $P = \dfrac{AM}{BM} = \dfrac{80}{8} = 10\,\text{m/s}$ which verifies the theoretical result.

Similarly, the rate of change of velocity with respect to time is called acceleration and is given by the gradient of the velocity–time graph at any point. In mathematical notation this is given by $\dfrac{dv}{dt}$.

Thus   Acceleration $a = \dfrac{dv}{dt}$

The above reasoning was applied to linear motion, but it could also have been used for angular motion. The essential difference is that distance s is replaced by angle turned through, θ rad.

Both sets of results are summarised in Fig. 13.7.

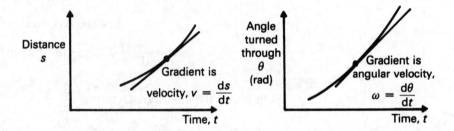

Linear motion   **Angular motion**

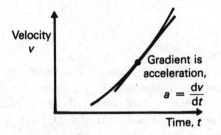
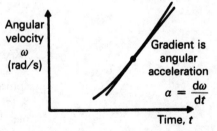

 Fig. 13.7

259

EXAMPLE 13.11

The research and development department of a company is developing a test rig to verify a theory for advanced dimensional analysis. A remotely controlled model has been programmed to move according to the equation $s = 2t^3 - 9t^2 + 12t + 6$ where s metres is the distance moved by the model at a time t seconds. We have been asked to find:

a) its velocity after 3 seconds,
b) its acceleration after 3 seconds,
c) when the velocity is zero.

We have
$$s = 2t^3 - 9t^2 + 12t + 6$$

$$\therefore \qquad v = \frac{ds}{dt} = 6t^2 - 18t + 12$$

and
$$a = \frac{dv}{dt} = 12t - 18$$

a) When $t = 3$, then the velocity

$$v = \frac{ds}{dt} = 6(3)^2 - 18(3) + 12 = 12 \text{ m/s}$$

b) When $t = 3$, then the acceleration

$$a = \frac{dy}{dt} = 12(3) - 18 = 18 \text{ m/s}^2$$

c) When the velocity is zero then $\frac{ds}{dt} = 0$.

That is
$$6t^2 - 18t + 12 = 0$$
$$\therefore \qquad t^2 - 3t + 2 = 0$$
$$\therefore \qquad (t-1)(t-2) = 0$$
$$\therefore \quad \text{either} \qquad t - 1 = 0 \quad \text{or} \quad t - 2 = 0$$
$$\therefore \quad \text{either} \qquad t = 1 \text{ second} \quad \text{or} \quad t = 2 \text{ seconds}$$

EXAMPLE 13.12

A displacement transducer has been fitted to a disc attached to a spindle, in order to measure the angle through which the spindle turns. After a period of observation of the machine tool being investigated, it has been established that the angular motion is governed by the relationship $\theta = 20 + 5t^2 - t^3$ where θ radians is the angle turned through at time t seconds.

We have been asked to calculate:

a) the angular velocity when $t = 2$ seconds.
b) the value of t when the angular deceleration is 4 rad/s^2.

We have
$$\theta = 20 + 5t^2 - t^3$$

$$\therefore \qquad \omega = \frac{d\theta}{dt} = 10t - 3t^2$$

and
$$\alpha = \frac{d\omega}{dt} = 10 - 6t$$

a) When $t = 2$, then the angular velocity

$$\omega = \frac{d\theta}{dt} = 10(2) - 3(2)^2 = 8 \text{ rad/s}$$

b) An angular deceleration of 4 rad/s² may be called an angular acceleration of -4 rad/s².

$\therefore$ when $\alpha = \dfrac{d\omega}{dt} = -4$ then $-4 = 10 - 6t$

                              or $t = 2.33$ seconds

Exercise 13.4

1) If $s = 10 + 50t - 2t^2$, where s metres is the distance travelled in t seconds by a body, what is the velocity of the body after 2 seconds?

2) If $v = 5 + 24t - 3t^2$ where v m/s is the velocity of a body at a time t seconds, what is the acceleration when $t = 3$?

3) A body moves s metres in t seconds where $s = t^3 - 3t^2 - 3t + 8$. Find:
   a) its velocity at the end of 3 seconds,
   b) when its velocity is zero,
   c) its acceleration at the end of 2 seconds,
   d) when its acceleration is zero.

4) A body moves s metres in t seconds, where $s = \dfrac{1}{t^2}$. Find the velocity and acceleration after 3 seconds.

5) The distance s metres travelled by a falling body starting from rest after a time t seconds is given by $s = 5t^2$. Find its velocity after 1 second and after 3 seconds.

6) The distance s metres moved by the end of a lever after a time t seconds is given by the formula $s = 6t^2$. Find the velocity of the end of the lever when it has moved a distance $\frac{1}{2}$ metre.

7) The angular displacement θ radians of the spoke of a wheel is given by the expression $\theta = \frac{1}{2}t^4 - t^3$ where t seconds is the time. Find:
   a) the angular velocity after 2 seconds.
   b) the angular acceleration after 3 seconds,
   c) when the angular acceleration is zero.

8) An angular displacement θ radians in time t seconds is given by the equation $\theta = \sin 3t$. Find:
   a) the angular velocity when $t = 1$ second,
   b) the smallest positive value of t for which the angular velocity is 2 rad/s,
   c) the angular acceleration when $t = 0.5$ seconds,
   d) The smallest positive value of t for which the angular acceleration is 9 rad/s².

9) A mass of 5000 kg moves along a straight line so that the distance s metres travelled in a time t seconds is given by $s = 3t^2 + 2t + 3$. If v m/s is its velocity and m kg is its mass, then its kinetic energy is given by the formula $\frac{1}{2}mv^2$. Find its kinetic energy at a time $t = 0.5$ seconds, remembering that the joule (J) is the unit of energy.

Turning points

When we look at the features of a mathematical curve we appreciate first the overall shape and layout. Other important features are where it cuts the reference axes and the turning points—these are especially of interest when using calculus, as you will see.

Consider the graph of $y = x^3 + 3x^2 - 9x + 6$.

To plot the graph we draw up a table in the usual way:

x	−5	−4	−3	−2	−1	0	1	2	3
$y = x^3 + 3x^2 - 9x + 6$	1	26	33	28	11	6	1	8	23

The graph is shown plotted in Fig. 13.8.

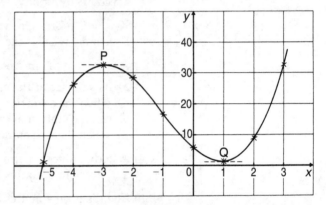

Fig. 13.8

The points P and Q are called turning points since the gradient of the tangent is zero at each of them. Also P is called a maximum and Q is called a minimum. It will be seen from the graph that the value of y at P is not the greatest value that y can have—nor the value of y at Q the least.

The terms maximum and minimum values apply only in the vicinity of the turning points and not to the values of y in general.

So at P, when $x = -3$, we have a maximum value of y of 33 and at Q, when $x = 1$, we have a minimum value of y of 1.

Maxima and minima

An important use of calculus is its application to finding maxima and minima, especially to a wide variety of engineering problems.

It is not always convenient to draw the full graph to find the turning points. At the turning points the gradient (slope) is zero, i.e. $\dfrac{dy}{dx} = 0$ in calculus notation (Fig. 13.9). As you will see this enables us to find the turning points.

Maximum or minimum?

It is usually necessary to determine if a turning point is a maximum or minimum. Fig. 13.10 shows how the gradient of a curve changes in the vicinity of a turning point.

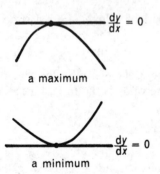

$\dfrac{dy}{dx} = 0$

a maximum

$\dfrac{dy}{dx} = 0$

a minimum

Fig. 13.9

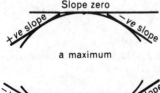

Slope zero

+ve slope   −ve slope

a maximum

−ve slope   +ve slope

Slope zero
a minimum

Fig. 13.10

EXAMPLE 13.13

Check the results for maximum and minimum values of y, found previously be a graphical means, given that $y = x^3 + 3x^2 - 9x + 6$.

We have
$$y = x^3 + 3x^2 - 9x + 6$$

$\therefore$
$$\frac{dy}{dx} = 3x^2 + 6x - 9$$

At a turning point
$$\frac{dy}{dx} = 0$$

$\therefore$
$$3x^2 + 6x - 9 = 0$$

We will now simplify by dividing through by 3, and solve the quadratic equation using factors:

thus $\qquad\qquad\qquad\qquad x^2 + 2x - 3 = 0$

or $\qquad\qquad\qquad\qquad (x-1)(x+3) = 0$

$\therefore$ either $\qquad\qquad\qquad (x-1) = 0$ or $x + 3 = 0$

so either $\qquad\qquad\qquad\qquad x = 1$ or $\qquad x = -3$

and these confirm the turning point positions shown in Fig. 13.8.

Test for maximum or minimum

At the turning point where $x = 1$, we know that $\frac{dy}{dx} = 0$, i.e. zero slope, and using a value of x slightly less than 1, say $x = 0.5$, gives $\frac{dy}{dx} = 3(0.5)^2 + 6(0.5) - 9 = -5.25$, i.e. negative slope, and using a value of x slightly greater than 1, say $x = 1.5$, gives $\frac{dy}{dx} = 3(1.5)^2 + 6(1.5) - 9 = +6.75$, i.e. positive slope.

These results are best shown by means of a diagram (Fig. 13.11) which indicates clearly that when $x = 1$ we have a minimum.

The minimum value of y may be found by substituting $x = 1$ into the given equation. Hence

$$y_{min} = (1)^3 + 3(1)^2 - 9(1) + 6 = 1$$

Similarly for the turning point at $x = -3$, at values of $x = -3.5$ (slightly below -3) and $x = -2.5$ (slightly above -3) we obtain the result shown in Fig. 13.12.

Fig. 13.12 indicates that when $x = -3$ we have a maximum.

The maximum value of y may be found by substituting $x = -3$ into the given equation. Hence

$$y_{max} = (-3)^3 + 3(-3)^2 - 9(-3) + 6 = 33$$

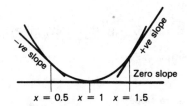

$x = 0.5 \quad x = 1 \quad x = 1.5$

Fig. 13.11

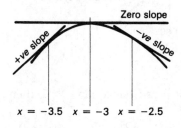

$x = -3.5 \quad x = -3 \quad x = -2.5$

Fig. 13.12

Applications in technology

In order to find maxima and minima using calculus we need an equation connecting the quantity for which a maximum or minimum is required in terms of another variable. It may well be necessary to form this equation and it may help to draw a diagram representing the problem.

EXAMPLE 13.14

An electric current is represented by $i = 5 \sin \theta$ where θ is in radians. Find the maximum value of i between $\theta = 0$ and π radians.

We have $$i = 5 \sin \theta$$

then $$\frac{di}{d\theta} = 5 \cos \theta$$

At a turning point $$\frac{di}{d\theta} = 0$$

i.e. $$5 \cos \theta = 0$$

or $$\cos \theta = 0$$

giving $$\theta = \frac{\pi}{2} \text{ rad}$$

This is the only solution between 0 and π radians.

Test for maximum or minimum

At the turning point where $\theta = \frac{\pi}{2} = 1.57$ rad we know that $\frac{di}{d\theta} = 0$, i.e. zero slope

and using a value of θ slightly less than 1.57, say $\theta = 1.50$, gives $\frac{di}{d\theta} = 5 \cos 1.50 = 0.35$, i.e. positive slope

and using a value of θ slightly greater than 1.57, say $\theta = 1.60$, gives $\frac{di}{d\theta} = 5 \cos 1.60 = -0.15$, i.e. negative slope.

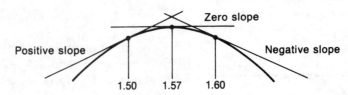

Fig. 13.13

Fig. 13.13 indicates a maximum at $\theta = \frac{\pi}{2}$ rad, since $\sin \frac{\pi}{2} = 1$ then the maximum value of current $i_{max} = 5 \sin \frac{\pi}{2} = 5$

EXAMPLE 13.15

A rectangular sheet of metal 360 mm by 240 mm has four equal squares cut out at the corners. The sides are then turned up to form a rectangular box. Find the length of the sides of the squares cut out so that the volume of the box may be as great as possible and find this maximum volume.

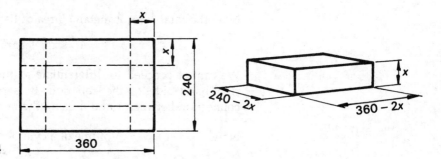

Fig. 13.14

Let the length of the side of each cut away square be x mm as shown in Fig. 13.14.

Hence the volume is
$$V = x(240 - 2x)(360 - 2x)$$
$$= 4x^3 - 1200x^2 + 86\,400x$$

$\therefore$
$$\frac{dV}{dx} = 12x^2 - 2400x + 86\,400$$

At a turning point
$$\frac{dV}{dx} = 0$$

$\therefore$
$$12x^2 - 2400x + 86\,400 = 0$$

or
$$x^2 - 200x + 7200 = 0 \quad \text{by dividing through by 12}$$

Now this is a quadratic equation which does not factorise easily so we will have to solve using the formula for the standard quadratic
$$ax^2 + bx + c = 0 \quad \text{giving} \quad x = \frac{-b \pm \sqrt{b^2 - 4ac}}{2a}$$

Hence the solution of our equation is

$$x = \frac{-(-200) \pm \sqrt{(-200)^2 - 4 \times 1 \times 7200}}{2 \times 1}$$

$\therefore$ either $\qquad\qquad x = 152.9 \quad \text{or} \quad x = 47.1$

However, from the physical sizes of the sheet, it is not possible for x to be 152.9 mm (since one side is only 240 mm long) so we reject this solution. Hence $x = 47.1$ mm.

We will leave you to check that the turning point at $x = 47.1$ is a maximum.

It only remains to find the maximum volume by substituting $x = 47.1$ into the equation for V. Therefore

$$V_{max} = 47.1(240 - 2 \times 47.1)(360 - 2 \times 47.1)$$
$$= 1.825 \times 10^6 \text{ mm}^3$$

EXAMPLE 13.16

A cylinder with an open top has a capacity of 2 m³ and is made from sheet metal. Neglecting any overlaps at the joints find the dimensions of the cylinder so that the amount of sheet steel used is a minimum.

Let the height of the cylinder be h metres and the radius of the base be r metres as shown in Fig. 13.15.

Now the total area of metal = area of base + area of curved side

$$A = \pi r^2 + 2\pi rh$$

We cannot proceed to differentiate as there are two variables on the right-hand side of the equation. It is possible, however, to find a connection between r and h using the fact that the volume is 2 m³.

Now   Volume of a cylinder = $\pi r^2 h$

$\therefore$ $2 = \pi r^2 h$

from which $h = \dfrac{2}{\pi r^2}$

We may now substitute for h in the equation for A.

$\therefore$

$$A = \pi r^2 + 2\pi r\left(\frac{2}{\pi r^2}\right)$$

$$= \pi r^2 + \frac{4}{r}$$

$$= \pi r^2 + 4r^{-1}$$

$\therefore$

$$\frac{\mathrm{d}A}{\mathrm{d}r} = 2\pi r - 4r^{-2}$$

Now for a turning point $\dfrac{\mathrm{d}A}{\mathrm{d}r} = 0$

or $2\pi r - 4r^{-2} = 0$

$\therefore$ $2\pi r - \dfrac{4}{r^2} = 0$

$\therefore$ $2\pi r = \dfrac{4}{r^2}$

$\therefore$ $r^3 = \dfrac{2}{\pi} = 0.637$

$$r = \sqrt[3]{0.637} = 0.860$$

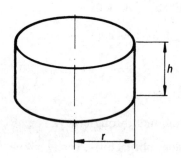

Fig. 13.15

Again we leave you to check that the turning point at $r = 0.86$ is a minimum.

Hence $r = 0.86$ m makes A a minimum as required.

We may find the corresponding value of h by substituting $r = 0.86$ into the equation found previously for h in terms of r.

$$h = \frac{2}{\pi(0.86)^2} = 0.86$$

Hence for the minimum amount of metal to be used the radius is 0.86 m and the height is also 0.86 m.

Exercise 13.5

1) Find the maximum and minimum values of:

   a) $y = 2x^3 - 3x^2 - 12x + 4$
   b) $y = x^3 - 3x^2 + 4$
   c) $y = 6x^2 + x^3$.

2) Given that $y = 60x + 3x^2 - 4x^3$, calculate:

   a) the gradient of the tangent to the curve of y at the point where $x = 1$,
   b) the value of x for which y has its maximum value,
   c) the value of x for which y has its minimum value.

3) Calculate the coordinates of the points on the curve

   $$y = x^3 - 3x^2 - 9x + 12$$

   at each of which the tangent to the curve is parallel to the x-axis.

4) A curve has the equation $y = 8 + 2x - x^2$. Find:

   a) the value of x for which the gradient of the curve is 6,
   b) the value of x which gives the maximum value of y,
   c) the maximum value of y.

5) The curve $y = 2x^2 + \dfrac{k}{x}$ has a gradient of 5 when $x = 2$. Calculate:

   a) the value of k,
   b) the minimum value of y.

6) Find the maximum and minimum values of the voltage v given by the expression $v = 3(\cos \theta) + 7$ between, and including, values of $\theta = 0$ and $\theta = 2\pi$ radians.

7) In an electric circuit the number of heat units, H, produced when a current, i, is flowing is given by $H = Ei - Ri^2$ where E is the e.m.f. and R is the resistance. Find the maximum heat that can be produced when $E = 8$ and $R = 2$.

8) The power, p watts, available from the rotor of an undershot waterwheel is given by $p = \dot{m}(v_j - u)u$. The water mass flow rate $\dot{m} = 7.85$ kg/s and the velocity of the vanes $v_j = 25$ m/s. Find the velocity, u, of the vanes for maximum power. Find also the value of the maximum power.

9) From a rectangular sheet of metal measuring 120 mm by 75 mm equal squares of side x are cut from each of the corners. The remaining flaps are then folded upwards to form an open box.

   Show that the volume of the box is given by

   $$V = 9000x - 390x^2 + 4x^3$$

   Find the value of x such that the volume is a maximum.

10) An open rectangular tank of height h metres with a square base of side x metres is to be constructed so that it has a capacity of 500 cubic metres. Show that the surface area of the four walls and the base will be $\dfrac{2000}{x} + x^2$ square metres. Find the value of x for this expression to be a minimum.

11) The volume of a cone is given by the formula $V = \frac{1}{3}\pi r^2 h$, where h is the height of the cone and r its radius. If $h = 6 - r$, calculate the value of r for which the volume is a maximum.

12) A box without a lid has a square base of side x mm nd rectangular sides of height h mm. It is made from $10\,800$ mm^2 of sheet metal of negligible thickness. Show that $h = \dfrac{10\,800 - x^2}{4x}$ and that the volume of the box is $(2700x - \frac{1}{4}x^3)$. Hence calculate the maximum volume of the box.

13) A cylindrical lemonade can made from thin metal has to hold 0.5 litres. Find its dimensions if the area of metal used is a minimum.

14) A cooling tank is to be made with the trapezoidal section as shown:

Its cross-sectional area is to be $300\,000$ mm^2. Show that the width of material needed to form, from one sheet, the bottom and folded-up sides is $w = \dfrac{300\,000}{h} + 1.828h$. Hence find the height h of the tank so that the width of material needed is a minimum.

15) A cylindrical cup is to be drawn from a disc of metal of 50 mm diameter. Assuming that the surface area of the cup is the same as that of the disc find the dimensions of the cup so that its volume is a maximum.

14

Areas under graphs

You are probably wondering why 'areas under graphs' justifies a chapter to itself. One important application is from a graph of force against distance moved — the area under this curve gives the work done. This can occur using indicator diagrams when testing engines, and as a rotary application in tests using rolling roads or dynamometers. There are a number of numerical rules, or methods, for finding areas. The first part of this chapter deals with these.

Finally you are introduced to the topic of integration. One of its important uses is for finding areas under graphs. If we have an equation for the graph, integration usually gives a rapid accurate answer. Occasionally it does not work and we have to fall back on the use of the rules — this is often called numerical integration.

Numerical methods for calculating areas and volumes

All the rules you will meet here basically do the same job. Sometimes one is more suitable than another — perhaps the easiest to use does not give the answer to the accuracy required? In addition to finding areas, these rules may be adapted to the assessment of volumes — another extremely useful application.

Areas

An irregular area is one whose boundary does not follow a definite pattern, e.g. the cross-section of a river.

In these cases practical measurements are made and the results plotted to give a graphical display.

Various numerical methods may then be used to find the area.

Mid-ordinate rule

Suppose we wish to find the area shown in Fig. 14.1. Let us divide the area into a number of vertical strips, each of equal width b.

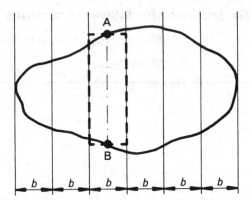

Fig. 14.1

Consider the 3rd strip, whose centre line is shown cutting the curved boundaries of the area at A and B respectively. Through A and B horizontal lines are drawn and these help to make the dotted rectangle shown. The rectangle has approximately the same area as that of the original 3rd strip, and this area will be $b \times AB$.

AB is called the mid-ordinate of the 3rd strip, as it is mid-way between the vertical sides of the strip.

To find the *whole area*, the areas of the other strips are found in a similar manner and then all are added together for the final result.

$\therefore$ Area = Width of strip × Sum of the mid-ordinates

> Mid-ordinate rule gives:
>
> Area = Width of strip × Sum of mid-ordinates

A useful practical tip to avoid measuring each separate mid-ordinate is to use a strip of paper and mark off along its edge successive mid-ordinate lengths, as shown in Fig. 14.2. The total area will then be found by measuring the whole length marked out (in the case shown this is HR) and multiplying by the strip width b.

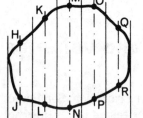

Fig. 14.2

EXAMPLE 14.1

Find the area under the curve $y = x^2 + 2$ between $x = 1$ and $x = 4$

The curve is sketched in Fig. 14.3. Taking 6 strips we may calculate the mid-ordinates.

x	1.25	1.75	2.25	2.75	3.25	3.75
y	3.56	5.06	7.06	9.56	12.56	16.06

Since the width of the strips $= \frac{1}{2}$,

the mid-ordinate rule gives

$$\text{Area} = \tfrac{1}{2} \times (3.56 + 5.06 + 7.06 + 9.56 + 12.56 + 16.06)$$
$$= \tfrac{1}{2} \times 53.86 = 26.93 \text{ square units}$$

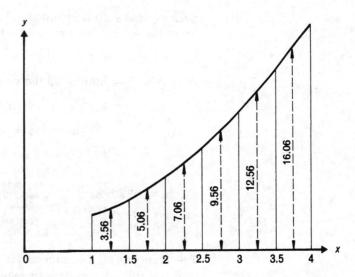

Fig. 14.3

It so happens that in this example it is possible to calculate an exact answer. How this is done need not concern us at this stage, but by comparing the exact answer with that obtained by the mid-ordinate rule we can see the size of the error.

$$\text{Exact answer} = 27 \text{ square units}$$

Approximate answer (using the mid-ordinate rule)

$$= 26.93 \text{ square units}$$

$$\therefore \quad \text{Error} = 0.07 \text{ square units}$$

$$\therefore \quad \text{Percentage error} = \frac{0.07}{27} \times 100 = 0.26\%$$

From the above it is clear that the mid-ordinate rule gives a good approximation to the correct answer.

Trapezoidal rule

Consider the area having boundary ABCD shown in Fig. 14.4.

The area is divided into a number of vertical strips of equal width b.

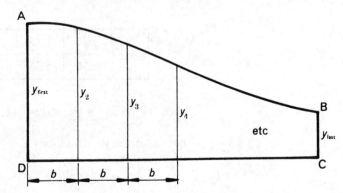

Fig. 14.4

Each vertical strip is assumed to be a trapezium. Hence the third strip, for example, will have an area $= b \times \frac{1}{2}(y_3 + y_4)$.

But

Area ABCD = Sum of all the vertical strips

$$= b \times \tfrac{1}{2}(y_{first} + y_2) + b \times \tfrac{1}{2}(y_2 + y_3) + b \times \tfrac{1}{2}(y_3 + y_4) + \ldots$$

$$= b\left[\tfrac{1}{2}y_{first} + \tfrac{1}{2}y_2 + \tfrac{1}{2}y_2 + \tfrac{1}{2}y_3 + \ldots + \tfrac{1}{2}y_{last}\right]$$

$$= b\left[\tfrac{1}{2}(y_{first} + y_{last}) + y_2 + y_3 + y_4 \ldots\right]$$

> **Area = Width of strip × $\left[\frac{1}{2}\right.$ (sum of the first and last ordinates)**
>
> **+ (sum of the remaining ordinates)$\left.\right]$**

The accuracy of the trapezoidal rule is similar to that of the mid-ordinate rule. A comparison may be made by solving Example 14.1 using the trapezoidal rule as in Example 14.2.

EXAMPLE 14.2

Find the area under the curve $y = x^2 + 2$ between $x = 1$ and $x = 4$

The curve is sketched in Fig. 14.5, the lengths of the ordinates having been calculated.

Since the width of the strips $= \frac{1}{2}$, the trapezoidal rule gives

$$\text{Area} = \tfrac{1}{2} \times [\tfrac{1}{2}(3 + 18) + 4.25 + 6 + 8.25 + 11 + 14.25]$$

$$= \tfrac{1}{2} \times [10.5 + 43.75]$$

$$= 27.13 \text{ square units}$$

The exact answer is 27 square units and therefore

$$\text{Percentage error} = \frac{27.13 - 27}{27} \times 100 = 0.48\%$$

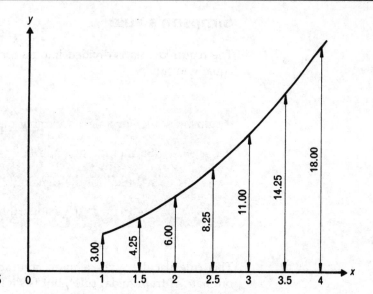

Fig. 14.5

EXAMPLE 14.3

The table gives the values of a force required to pull a trolley when measured at various distances from a fixed point in the direction of the force.

F (N)	51	49	45	37	26	15	10
s (m)	0	1	2	3	4	5	6

Calculate the total work done by this force.

The force–distance graph is plotted as shown in Fig. 14.6. The required work done is given by the shaded area under the curve. This may be found by dividing the area into strips and using the trapezoidal rule.

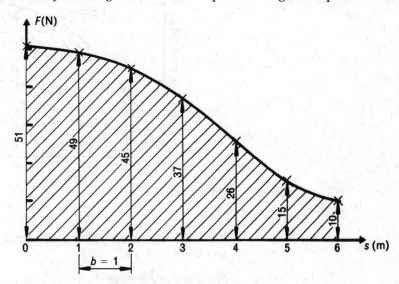

Fig. 14.6

Therefore   Area $= 1[\frac{1}{2}(51 + 10) + (49 + 45 + 37 + 26 + 15)] = 203$

Thus   Work done $= 203 \, \text{Nm} = 203 \, \text{J}$ (since joule $=$ newton $\times$ metre)

Simpson's rule

The required area is divided into an *even* number of vertical strips of equal width b.

> Simpson's rule, for an even number of strips, gives:
>
> $$\text{Area} = \frac{b}{3}\,[(\text{Sum of the first and last ordinates})$$
>
> $$+\ 2\ (\text{Sum of the remaining odd ordinates})$$
>
> $$+\ 4\ (\text{Sum of the even ordinates})\,]$$

This rule usually gives a more accurate result than either the mid-ordinate of trapezoidal rules, but is slightly more complicated to use.

Note. **There must be an *even* number of strips.**

EXAMPLE 14.4

A compressor unit has been causing problems in service by over-heating. One possible way of rectifying this is to increase the surface area of the cooling fins. The first step, therefore, is to calculate the surface area of the existing fins. The profile of a fin is shown in Fig. 14.7 all dimensions being in millimetres.

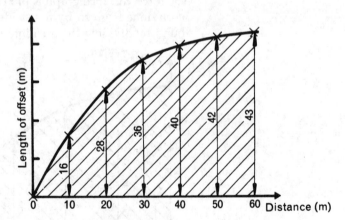

Fig. 14.7

The area under the graph gives the area of the fin and it is shown shaded in Fig. 14.7.

Simpson's rule gives

$$\text{Area} = \tfrac{10}{3}[(0+43)+2(28+40)+4(16+36+42)] = 1850$$

Therefore area of the fin is 1850 mm^2

Exercise 14.1

1) The area under a velocity–time graph gives the distance travelled. During a test on a motor vehicle its velocity was measured at 5 second intervals and the results were as follows:

Velocity (m/s)	0	4	15	23	17	9	3	0
Time (s)	0	5	10	15	20	25	30	35

Plot the velocity–time curve and find the distance travelled overall. Find also the distance travelled between 5 and 25 seconds.

2) An electric current is given by $4 \sin \theta$ amperes. Find its mean value over half a cycle.

Explanation: the mean value of the current is given by the area under the curve divided by the length of the base, provided that the base is measured in radian units. So plot the waveform over half a cycle (0° to 180°, or 0 to π radians), divide it into suitable strips and the rest should be reasonably straightforward.

3) A gas expands according to the law $pv^{1.3} = 175$ where p kN m^{-2} is the pressure and v m^3 is the volume. Find the work done in kJ units when the gas expands from 1.1 m^3 to 1.9 m^3.

Explanation: the work done by the gas as it expands is found from the area under the pressure–volume curve. So first transpose the law to make p the subject, and plot a graph of p vertically against v horizontally. Then use one of the area laws to find the work done.

4) One of the products from the firm for whom you are a development engineer is a meter for measuring the rate of flow of water in rivers. In order to check the quantity of water flowing at a particular location the local water board have furnished you with a set of soundings taken across the river cross-section. Your first task is to calculate the cross-sectional area. Details of the soundings are given in Fig. 14.8.

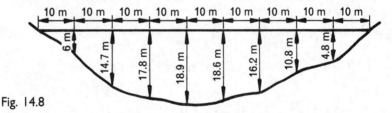

Fig. 14.8

5) We are involved with the design of a large plant for the chemical industry. One of the distillation cylinders is ventilated by a fabricated steel ducting with details as shown in the table:

Distance from one end (m)	0	1	2	3	4	5	6
Cross-sectional area (m^2)	5.1	4.1	3.4	2.7	2.2	1.8	1.3

You wish to find the volume encompassed by the ducting. This may be achieved by taking the values of the cross-sectional areas as ordinates, and using one of the rules which, in this case, will provide the required volume.

Integration

One way to approach integration is to think of it as the reverse of differentiation — the rules we develop will be 'backwards' versions of those we used for differentiation. However, the basic idea comes from areas, whereas for differentiation it came from gradients.

Integration as the inverse of differentiation

We have previously discovered how to obtain the differential coefficients of various functions. Our objective in this section is to find out how to reverse the process. That is, being given the differential coefficient of a function we try to discover the original function.

if $$y = \frac{x^4}{4}$$

then $$\frac{dy}{dx} = x^3$$

or we may write $$dy = x^3 \, dx$$

The expression $x^3 \, dx$ is called the differential of $\frac{x^4}{4}$.

Reversing the process of differentiation is called *integration*.

It is indicated by using the integration sign $\int$ in front of the differential.

Thus, if: $$dy = x^3 \, dx$$

then reversing the process $$y = \int x^3 \, dx = \frac{x^4}{4}$$

Similarly if $$y = \frac{x^5}{5}$$

then $$\frac{dy}{dx} = x^4$$

or $$dy = x^4 \, dx$$

and reversing the process $$y = \int x^4 \, dx = \frac{x^5}{5}$$

Also, if $$y = \frac{x^{n+1}}{n+1}$$

then $$\frac{dy}{dx} = x^n$$

or $$dy = x^n \, dx$$

from which $$y = \int x^n \, dx = \frac{x^{n+1}}{n+1}$$

Now $\dfrac{x^{n+1}}{n+1}$ is called the integral of $x^n \, dx$.

This rule applies to all indices whether positive, negative or fractional except for $\int x^{-1} \, dx = \int \dfrac{1}{x} \, dx$

Since $\dfrac{d}{dx}(\log_e x) = \dfrac{1}{x}$ the reverse of this gives:

$$\int \frac{1}{x} \, dx = \log_e x$$

Since $\dfrac{d}{dt}\sin(\omega t + \alpha) = \cos(\omega t + \alpha)$ the reverse of this gives:

$$\int \cos(\omega t + \alpha) \, dt = \frac{1}{\omega}\sin(\omega t + \alpha)$$

Similarly

$$\int \sin(\omega t + \alpha) \, dt = -\frac{1}{\omega}\cos(\omega t + \alpha)$$

Since $\dfrac{d}{dx}e^{ax} = a\,e^{ax}$ the reverse of this gives:

$$\int e^{ax} \, dx = \frac{1}{a}\,e^{ax}$$

Summarising:

$$\int x^n \, dx = \frac{x^{n+1}}{n+1}$$

$$\int \cos(\omega t + \alpha) \, dt = \frac{1}{\omega}\sin(\omega t + \alpha)$$

$$\int \sin(\omega t + \alpha) \, dt = -\frac{1}{\omega}\cos(\omega t + \alpha)$$

$$\int \frac{1}{x} \, dx = \log_e x$$

$$\int e^{ax} \, dx = \frac{1}{a}\,e^{ax}$$

The constant of integration

We know that the differential of $\frac{1}{2}x^2$ is $x\,dx$. Therefore if we are asked to integrate $x\,dx$, $\frac{1}{2}x^2$ is one answer; but it is not the only possible answer because $\frac{1}{2}x^2 + 2$, $\frac{1}{2}x^2 + 5$, $\frac{1}{2}x^2 + 19$, etc. are all expressions whose differential is $x\,dx$. The general expression for $\int x\,dx$ is therefore $\frac{1}{2}x^2 + c$, where c is a constant known as the constant of integration. Each time we integrate the constant of integration must be added.

EXAMPLE 14.5

a) $\displaystyle\int x^5\,dx = \frac{x^{5+1}}{5+1} + c = \frac{x^6}{6} + c$ b) $\displaystyle\int x\,dx = \frac{x^{1+1}}{1+1} + c = \frac{x^2}{2} + c$

c) $\displaystyle\int \sqrt{x}\,dx = \int x^{1/2}\,dx = \frac{x^{3/2}}{3/2} + c = \frac{2x^{3/2}}{3} + c$

d) $\displaystyle\int \frac{dx}{x^3} = \int x^{-3}\,dx = \frac{x^{-2}}{-2} + c = -\frac{1}{2x^2} + c$

e) $\displaystyle\int \cos 3\theta\,d\theta = \frac{1}{3}(\sin 3\theta) + c$

A constant coefficient may be taken outside the integral sign

EXAMPLE 14.6

a) $\displaystyle\int 6e^{2x}\,dx = 6\int e^{2x}\,dx = 6(\tfrac{1}{2}e^{2x}) + c = 3e^{2x} + c$

b) $\displaystyle\int 4\sin\theta\,d\theta = 4\int \sin\theta\,d\theta = 4(-\cos\theta) + c = -4\cos\theta + c$

The integral of a sum is the sum of their separate integrals

EXAMPLE 14.7

a) $\displaystyle\int (x^2 + x)\,dx$

Integrate each term separately:

$$\int x^2\,dx = \frac{x^3}{3} \qquad \text{and} \qquad \int x\,dx = \frac{x^2}{2}$$

Thus $$\int (x^2 + x)\,dx = \frac{x^3}{3} + \frac{x^2}{2} + c$$

b) $\displaystyle\int \left(\frac{1}{x} + 7\right) dx = (\log_e x) + 7x + c$

c) $\displaystyle\int (3\sin t - 5\cos t)\,dt = -3\cos t - 5\sin t + c$

d) $\displaystyle\int (e^{4u} + e^{-4u})\,du = \frac{1}{4}e^{4u} + \frac{1}{(-4)}e^{-4u} + c = \frac{1}{4}e^{4u} - \frac{1}{4}e^{-4u} + c$

Exercise 14.2

Integrate with respect to the variable in each example.

1) x^2

2) $\sin \theta$

3) $\sqrt{x}$

4) $\dfrac{1}{x^2}$

5) $\dfrac{1}{x}$

6) e^{8t}

7) $3 \cos 2\theta$

8) $5x^8 + e^x$

9) $\dfrac{1}{x} + x + 3$

10) $2 \cos \theta - \sin 3\theta$

11) $6 + 5x + \dfrac{1}{\sqrt{x}} + \dfrac{2}{x^2}$

12) $10 + \dfrac{1}{2}(e^{5u} + e^{-5u})$

Evaluating the constant of integration

The value of the constant of integration may be found provided a corresponding pair of values of x and y are known.

EXAMPLE 14.8

The gradient of the curve which passes through the point $(2, 3)$ is given by x^2. Find the equation of the curve.

We are given
$$\frac{dy}{dx} = x^2$$

$\therefore$
$$y = \int x^2 \, dx = \frac{x^3}{3} + c$$

We are also given that when $x = 2,\ y = 3$

Substituting these values in
$$y = \frac{x^3}{3} + c$$

we have
$$3 = \frac{2^3}{3} + c$$

$\therefore$
$$c = \frac{1}{3}$$

Hence the equation of the curve is

$$y = \frac{x^3}{3} + \frac{1}{3}$$

or
$$y = \frac{1}{3}(x^3 + 1)$$

Exercise 14.3

1) The gradient of the curve which passes through the point (2, 3) is given by x. Find the equation of the curve.

2) The gradient of the curve which passes through the point (3, 8) is given by $(x^2 + 3)$. Find the value of y when $x = 5$.

3) It is known that for a certain curve $\dfrac{dy}{dx} = 3 - 2x$ and the curve cuts the x-axis where $x = 5$. Express y in terms of x. State the length of the intercept on the y-axis.

4) Find the equation of the curve which passes through the point (1, 4) and is such that $\dfrac{ds}{dt} = e^{3t}$.

5) If $\dfrac{dp}{dt} = \dfrac{1}{t}$ find p in terms of t given that $p = 3$ when $t = 2$.

6) The gradient of a curve is $ax + b$ at all points, where a and b are constants. Find the equation of the curve given that it passes through the points (0, 4) and (1, 3) and that the tangent at (1, 3) is parallel to the x-axis.

7) Find the equation of the curve which passes through the point (1, 2) and has the property of $\dfrac{dy}{dx} (\cos x)$.

8) A curve is such that $\dfrac{dy}{d\theta} = \cos \theta$, and also $y = 1$ when $\theta = \dfrac{\pi}{2}$ radians. Find the equation of the curve.

9) At any point on a curve $\dfrac{dy}{d\theta} = 3 \sin \theta$. Find the equation of the curve given that $y = 2$ when θ has a value equivalent to 25 degrees.

The definite integral

It has been shown that $\displaystyle\int x^n \, dx = \dfrac{x^{n+1}}{n+1} + c.$

Since the expression contains an arbitrary constant c, the value of which is not known, it is called an indefinite integral.

A definite integral has a specific numerical answer without an unknown constant. The notation for this definite integral is $\displaystyle\int_a^b x^n \, dx.$

a and b are called limits, a being the lower limit and b the upper limit. The method of evaluating a definite integral is shown in the following examples.

EXAMPLE 14.9

Find the value of $\int_2^3 x^2 \, dx$

$$\int_2^3 x^2 \, dx = \left[\frac{x^3}{3} + c \right]_2^3$$

$$= \left(\text{Value of } \frac{x^3}{3} + c \text{ when } x \text{ is put equal to 3} \right)$$

$$- \left(\text{Value of } \frac{x^3}{3} + c \text{ when } x \text{ is put equal to 2} \right)$$

$$= \left(\frac{3^3}{3} + c \right) - \left(\frac{2^3}{3} - c \right)$$

$$= \frac{27}{3} + c - \frac{8}{3} - c$$

$$= \frac{19}{3} = 6.33$$

Square brackets in integration

In integration the use of the square brackets, as in the above solution, has a specific meaning, that is *the integration of each term has been completed and the next step is to substitute the values of the limits for x.*

We should also note that the constant c cancelled out. This will always happen and in solving definite integrals it is usual to omit c as shown in the next example.

EXAMPLE 14.10

Find the value of $\int_1^2 (3x^2 - 2x + 5) \, dx$

$$\int_1^2 (3x^2 - 2x + 5) = \left[x^3 - x^2 + 5x \right]_1^2$$

$$= (2^3 - 2^2 + 5 \times 2) - (1^3 - 1^2 + 5 \times 1)$$

$$= 14 - 5 = 9$$

EXAMPLE 14.11

Find the value of $\int_0^{\pi/2} \sin \theta \, d\theta$

$$\int_0^{\pi/2} \sin \theta \, d\theta = \left[-\cos \theta \right]_0^{\pi/2} = \left(-\cos \frac{\pi}{2} \right) - (-\cos 0)$$

$$= 0 - (-1) = 1$$

Exercise 14.4

Evaluate the following definite integrals:

1) $\int_1^2 x^2 \, dx$

2) $\int_2^3 (2x + 3) \, dx$

3) $\int_3^5 \frac{1}{v} \, dv$

4) $\int_0^\pi (3 \cos \theta) \, d\theta$

5) $\int_1^2 (7 + e^{2t}) \, dt$

6) $\int_0^2 \sqrt{x} \, dx$

Area under a graph

The application of integration to finding of areas under graphs is extremely important in technology.

You have already met some instances in engineering where it was necessary to calculate areas under graphs, and we had to use area rules which gave approximate results.

Now, providing we have the equation of a graph, we may find the area underneath the graph using integration, and obtain a much more accurate value.

Suppose that we wish to find the shaded area shown in Fig. 14.9. P, whose coordinates are (x, y), is a point on the curve.

Let us now draw, below P, a vertical strip whose width δx is very small. Since the width of the strip is very small we may consider the strip to be a rectangle with height y. Hence the area of the strip is approximately $y \times \delta x$. Such a strip is called an elementary strip and we will consider that the shaded area is made up from many elementary strips. Hence the required area is the sum of all the elementary strip areas between the values $x = a$ and $x = b$. In mathematical notation this may be stated as:

$$\text{Area} = \sum_{x=a}^{x=b} y \times \delta x \qquad \text{approximately}$$

The process of integration may be considered to sum up an infinite number of elementary strips and hence gives an exact result.

$$\therefore \qquad \text{Area} = \int_a^b y \, dx \qquad \text{exactly}$$

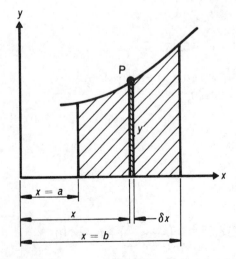

Fig. 14.9

EXAMPLE 14.12

A voltage is given by $v = R \sin \theta$. What will be its mean value over a half wave?

The mean value is given by the area under a graph divided by the length of the base of the area.

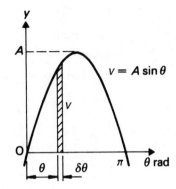

Fig. 14.10

Now a complete cycle, or full wave as it is sometimes called, of a sine waveform occurs over a range of 0° to 360° (or 0 to 2π radians), so a half wave goes from $\theta = 0$ rad to $\theta = \pi$ rad as shown in Fig. 14.10.

$$\text{Area of elementary strip} = v \, \delta\theta$$

$\therefore \quad$
$$\text{Total area under curve} = \sum_{\theta=0}^{\theta=\pi} v \, \delta\theta \qquad \text{approximately}$$

$$= \int_0^\pi v \, \mathrm{d}\theta \qquad \text{exactly}$$

$$= \int_0^\pi R \sin \theta \, \mathrm{d}\theta$$

$$= R \int_0^\pi \sin \theta \, \mathrm{d}\theta$$

$$= R[-\cos]_0^\pi$$

$$= R\{(-\cos \pi) - (-\cos 0)\}$$

$$= R\{-(-1) - (-1)\}$$

$$= 2R \text{ square units}$$

Now
$$\text{Mean value} = \frac{\text{Area under curve}}{\text{Length of base}}$$

$$= \frac{2R}{\pi}$$

$$= 0.637R$$

So the mean value of a sine waveform over half a cycle is 0.637 times the amplitude — an important result.

EXAMPLE 14.13

A gas expands according to the law $pv = c$, where p is the pressure, v is the volume and c is a constant. We know that the pressure is $250 \, \text{N m}^{-2}$ when the volume is $5 \, \text{m}^3$. We need to find the work done by gas as it expands from $4 \, \text{m}^3$ to $8 \, \text{m}^3$.

First we may find the value of the constant c by substituting the values of $p = 250$ when $v = 5$ into the stated law

and obtain $\qquad\qquad\qquad\qquad\qquad 250 \times 5 = c$

giving $\qquad\qquad\qquad\qquad\qquad\qquad\quad c = 1250$

and the law becomes $\qquad\qquad\qquad\qquad pv = 1250$

from which $\qquad\qquad\qquad\qquad\qquad\quad p = \dfrac{1250}{v}$

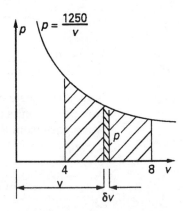

Fig. 14.11

Now the work done by expanding gas is given by the area under the graph of pressure p plotted against volume v, which from Fig. 14.11 is given by

$$\text{Work done} = \sum_{x=4}^{x=8} p \, \delta v \qquad \text{approximately}$$

$$= \int_{4}^{8} \frac{1250}{v} \, dv \qquad \text{exactly}$$

$$= 1250 \int_{4}^{8} \frac{1}{v} \, dv$$

$$= 1250 [\log_e v]_{4}^{8}$$

$$= 1250(\log_e 8 - \log_e 4)$$

$$= 866 \, \text{kJ} \qquad \text{(since the units of } p \text{ are N m}^{-2}$$

$$\text{and the units of } v \text{ are m}^3)$$

Exercise 14.5

1) A coil spring has a stiffness of 16 N/mm, and has been chosen for use in a safety valve. You have been requested to give advice on the suitability of the spring, and one of your first tasks is to find the work done when the spring is compressed over its working range, from a length of 70 mm to 50 mm.

   Information: the stiffness, or rate, S of a spring may be found from the force F required to compress, or extend, the spring through a distance x. So $S = \dfrac{F}{x}$ from which $F = Sx$ and if you sketch the graph of F against x you will be able to set up an integral and hence find the work done in an appropriate SI preferred unit.

2) The velocity of an automatic chuck for moving the wire through a nail-making machine may be found from the expression $v = 0.1t(1 - t)$ where v is the velocity in m s^{-1} and t is the time in seconds. You have been asked to calculate the distance moved by the chuck between the two times when the velocity is zero.

   Information: the area under a velocity–time graph gives the distance travelled. Sketch the graph, set up an integral and you are on your way!

3) A recoil mechanism fixed to the spindle of a light machine tool is under scrutiny as failures have been occurring. One of the problems is to find how much energy the spiral spring, in the mechanism, can absorb when it is wound up through half a revolution of the spindle from the unwound-up datum. The spring manufacturers have furnished you with an expression $T = 5 + 0.9\theta$ where T N m is the torque required to wind up the spring through an angle θ radians. You have offered to find the energy value, as you know that the energy absorbed is the work done in winding up the spring — also that work done in torsion, in joule units, is the product of torque in N m units and angle turned through in radians. So a sketch graph of T against θ is required and you will be able to set up an integral to find the area under the T–θ graph.

4) In Exercise 14.1, Question 2 you were asked to find the mean value, over half a cycle, of a current $4 \sin \theta$ amperes. You are now in a position to confirm the result using integration. Sketch the curve of $i = 4 \sin \theta$ and follow the usual procedure.

5) In Exercise 14.1, Question 3 you were asked to find the work done by a gas expanding from 1.1 m^3 to 1.9 m^3 if the gas obeys a law $pv^{1.3} = 175$ where p kN m^{-2} is the pressure and v m^3 is the volume. Sketch a suitable graph and check the result by integration.

Portfolio problems 3

Proving ring

Fig. P3.1

1 You are an engineer working in a standards laboratory. A customer has just brought in a proving ring which, for annual certification, has to be recalibrated. The ring is no. PR 760932 owned by Mako Ltd.

Background A typical piece of measuring equipment is a proving ring (Fig. P3.1), which comprises a steel ring with a flat foot and a flat hat to support compressive loads. The dial gauge is used to measure the deflection of the ring for various loads. A typical ring is 250 mm diameter with a rectangular section 30 mm by 25 mm. Before the ring may be used to measure loads it has to be calibrated. This involves loading the test machine steadily and taking dial gauge readings corresponding to particular loads. A calibration graph then has to be drawn.

You have mounted the ring into the compression test machine and taken readings at 5 kN intervals, obtaining the following results:

Load (kN)	5	10	15	20	25	30	35	40	45	50
Dial gauge reading	35	66	97	130	160	190	224	254	285	317

Plot the graph of load vertically against deflection horizontally, as large as conveniently possible on the graph paper. You should show neatly your plotted points and draw the best curve or straight line. Finally be careful to put all relevant details on the top of the graph paper—an anonymous graph is no good to anyone! And remember your name, company, date of test and details of your test equipment, i.e. the machine in this case.

2 An establishment in the nuclear power industry is involved in the design and manufacture of thermodynamic machinery. There has been a problem with a constant pressure process on a certain gas. A formula is used which relates the specific entropy s with the absolute temperature T and is $T = 20\,e^{2.8s}$. For values of s from 1.00 to 1.25 (plotted horizontally), you have been asked to plot a graph of the function, and then obtain the values of T when $s = 1.10$, and of s when $T = 475$. A value of the area under the curve between s values of 1.00 and 2.00 has also been requested. Use one of the numerical rules which you consider suitable, and confirm your result by using integral calculus.

3 You have just joined a firm of consultant engineers and have been called in by a company who have problems with a small flat belt drive. They are querying the figure for the coefficient of friction, μ, between the belt and the pulley wheel as given to them by the belt suppliers. So you take the pulley wheel and belt sample provided and obtain permission to use a laboratory rig at the local college of technology. The rig (Fig. P3.2) comprises a pulley wheel, which you find can conveniently be temporarily replaced by your wheel, on the same spindle as a swinging arm which can be pegged in different positions. This enables you to vary the angle of lap, θ, which is the angle over which contact is made between belt and pulley. As you turn the pulley clockwise you note the spring balance reading F_1 N and the angle θ, and obtain the following results:

F_1 newtons	7.4	11.1	15.7	21.8	31.0	43.1	60.9
θ degrees	60	90	120	150	180	210	240

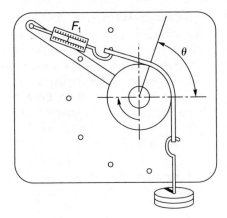

Fig. P3.2

The expression you will use is $F_1/F_2 = e^{\mu\theta}$ where θ is in radians. By reducing this equation to a straight line form (logs may be needed), find a value for the coefficient of friction μ. Not required in the calculations here is F_2, the tension in the slack side of the belt — in this case the hanger load.

4   You work in a part of a research and development section of a company specialising in the manufacture of electrical measuring equipment for the aircraft industry. You have been investigating how a current builds up in an electrical inductive circuit and have established that the expression $i = 2(1 - e^{10t})$ is a suitable expression relating the current i A to the time t s. You have been asked to find the following facts:

a) how long it takes for the current to increase from 1 A to 1.6 A
b) the rate of growth initially and at times 0.1 and 0.2 seconds.

You should first obtain solutions theoretically, assisted by the use of calculus, and then confirm the values obtained by graphical means (i.e. drawing a graph of the given equation, etc.).

5   You have been seconded by your firm, which makes containers and other associated equipment for the chemical industry, to spend some time in the laboratories of a pharmaceutical company. The idea is that you should observe their processes and activities in order to help in the design of new equipment they wish to order. You have shown interest in how they study the growth of a bacterial population, especially when you noticed that the equation for this is similar to those used in electrical and mechanical engineering. This relationship is $(10^{-10})e^{1.35t} = m$, where m grams is the mass of a bacterial population at a time t hours from the beginning of growth. So you plotted a graph of m vertically against time t and the following information was requested from you:

a) the mass of the population at the beginning of growth,
b) the time when the population has doubled from its initial value.

The laboratory assistants were grateful for your help as they were always hard pressed for time. However, they were quite impressed when you showed them how to get these numerical results quickly with the aid of differential calculus—show details of how you did this.

6   Concbridge Ltd are a company manufacturing reinforced concrete beams, specialising in prestressed steel varieties. You are employed in the standards laboratory of the firm and as such are responsible for the regular checking of the concrete mix, water content and other ongoing items, which includes checking the tensions in the reinforcing bars prior to casting. Although these are measured at the ends by strain gauges, where the bars are secured, periodic checks are made using a wire tension meter clamped on to a wire bar. A handle is rotated and the wire is pulled sideways to set a position where the reading of the handle position corresponds to the tension in the wire bar. The meter is calibrated by fitting it to a wire bar mounted in jaws on a tensile testing machine. Then tensile load readings are recorded against corresponding meter readings. You have obtained these results:

Load on wire (kN)	202	269	349	429	492	553	621	696	756	813	906	966	1020	1070
Meter reading	15	20	25	30	35	40	45	50	55	60	65	70	75	80

Plot a suitable calibration graph, with load vertically. Don't forget to label the axes and to put details of the tension meter (no. EU 5732A41BD) and the testing machine together with your name and date of test. Finally, use the calibration graph to find the tensions in a wire bar if the meter readings are 481 and 630.

7   A problem concerning the compression of a close coiled helical spring has been brought in for your attention. The spring. used as a buffer in an automatic machine tool, has an initial compression of 0.2 m and is then compressed a further 0.15 m. The spring rate (stiffness) is 0.2 kN/mm. Find the work done in compressing the spring the further 0.15 m.

Task 1: your first thought should be that the spring will obey Hooke's law which stages that 'Force is proportional to compression'. This, in equation form using symbols, is $F = Sx$ in which the force is F N, the compresion x mm and the spring rate S N/mm. Now the work done is the area under the graph of F (plotted vertically) against x, and will be given by the integral between the limits of $x = 0.2$ and $x = 0.35$. Now substitute for P in terms of x using the Hooke's law equation, integrate, substitute the values of the limits and the result is the work done.

Task 2: check your result graphically by plotting a suitable graph and using a suitable numerical rule for finding the area.

8   It has been suggested to you by the publicity and sales department of the company for which you work that you are just the person to prepare a page for a technical publication. This is a typical back-handed compliment, but you are not too worried when they explain that this page will demonstrate how to find the mean value of a half cycle of a trigonometrical sine waveform.

On asking around you find that a mean value is obtained from:

$$\text{mean value} = \frac{\text{area under a graph}}{\text{base length}}$$

with a proviso that when used with trigonometric applications the base length is in radians, not degrees. So you should first draw a sine wave for a half cycle and then use, say, Simpson's rule to find the area and thus the mean value. Remember it should be reasonably accurate and neat since it is for 'public scrutiny', and should include any words of explanation you consider necessary to clarify your presentation. And now the bombshell! You have also been asked to prepare an alternative script using a sketch graph and integration, in case a solution by calculus is preferred.

9   You have been asked to help with the design of a pressurised container to be used for a spray type of household furniture polish. It is to be made from thin sheet metal and have a capacity of 300 ml. Its shape is to be cylindrical but the ends are hemispherical—the top is convex and the bottom is concave. It has been suggested that you should ignore the effect of the thickness of the metal itself and also of any overlaps at the joints. As this is a competitive market, every effort is being made to keep the cost of the product as low as possible. To this end your task is to find the dimensions of the can so that the area of metal used in its construction is a minimum.

10 The research and development department of the petroleum company for which you work is faced with the following problem. It is known that a particular gas expands according to the law $pv^n = C$ and you have been asked to find the work done when the gas expands from $1\,\text{m}^3$ to $2\,\text{m}^3$. You have also been furnished with the following pairs of values which, you gather, have come from the experimental section:

Volume, v (m³)	0.3	0.6	0.9	1.2	1.5	1.8	2.1
Pressure, p (kN m⁻²)	1079.1	408.9	231.8	154.9	113.4	87.8	70.8

Your problem may be divided into two parts.

First: rearrange the given equation into logarithmic form and then plot a graph on suitable logarithmic paper. This will enable you to find the values of the constants C and n.

Second: in the case of an expanding gas the work done in expansion is given by 'pressure multiplied by volume' or, as an integral, . So you have to obtain p in terms of v from the gas law, substitute for p in terms of v in the integral, integrate the resulting expression and put in the values of the limits. Thus you will obtain the work done in kilojoules (kJ).

11 One of the senior persons with whom you work has been asked to give a lecture and he has requested that you should prepare two diagrams which can be made into slides for an overhead projector. The first diagram comprises one complete cycle of the waveforms $2\sin\theta$, $\sin\theta$, $\sin\dfrac{\theta}{2}$ and $\sin 2\theta$ all plotted on the same axes. The second diagram comprises one complete cycle of each of three waveforms, namely, $\cos\theta$, $\cos\left(\theta+\dfrac{\pi}{3}\right)$ and $\cos\left(\theta-\dfrac{\pi}{6}\right)$. Again these should be plotted on the same axes. For this second diagram you may find it clearer if you extend the plot of the last two curves outside the range of 0° to 360°.

Remember these graphs must look neat and tidy as would be expected from a professional engineer. Each curve should be labelled, and in the second diagram you should find some way of indicating the angles of lead and lag between the curves, stating clearly which lags, or leads, which.

Finally, remember that trigonometrical curves go on for ever and extend infinitely in either direction—show this by carefully putting 'tails' on all the curves and not ending them abruptly on the horizontal axis.

12 You and your colleagues have been involved in a laboratory study concerned with dynamics, which involves the study of distance s metres, time t seconds, velocity v metres per second, acceleration a metres per second per second and mass m kilograms. You are trying to predict what will happen if a mass of 0.231 kg moves according to the relationship given by $s = t^3 - 2t^2 + t + 3$.

You have been asked to work out the following:

a) the velocities when $t = 0$ and $t = 5$ seconds
b) the times when the velocity is zero
c) the acceleration when $t = 6$ seconds
d) the time when the velocity is either a maximum or a minimum
e) the value of the velocity found in part d) and whether it is a maximum or minimum
f) the kinetic energy of the mass at the instant when $t = 7$ seconds
g) the work done in slowing the mass down from 300 metres per second to 100 metres per second
h) the average force needed in part g).

Portfolio assignment 3

Part three of this book has basically looked at the graphical approach to problems. Apart from being able to calculate precise answers to problems we have come across ways of finding approximate results using numerical methods—just such an application follows in this assignment. When completed this work is eminently suitable for inclusion in your portfolio.

The engine test

Introductory background

Interest has recently been shown in an old engine which your company used to produce. It was a slow-running four-stroke petrol engine suitable for a light wagon. Although not as efficient as many present day engines, it was reliable and easy to maintain which makes it suitable for a developing country. An extensive search revealed the details of an engine test in the thermodynamics laboratory with all the essential data present but no calculations of any results.

Old test information

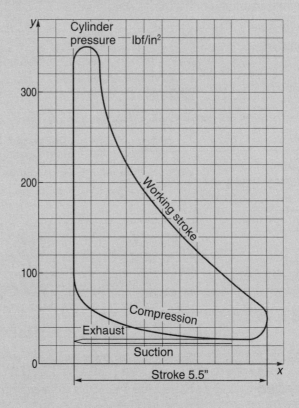

Fig. P3.3

Piston diameter	4 in
Stroke	5.5 in
Revolutions per minute	1500
Maximum (probable) explosion pressure	350 lbf/in^2
Compression ratio	4:1
Mass of reciprocating parts, per cylinder	4 lb

Explanation and theory

The indicator diagram shown in Fig. P3.3 is a graph of the pressure in one of the cylinders plotted against the position of the piston, and this was automatically recorded during the test.

Now the general expression for work done is force × distance moved, but the indicator diagram gives pressure vertically instead of force. We could use the fact that   force = pressure × area on which it acts and relabel the vertical axis in force units. But a better approach is to find the required area in pressure × distance units, and then make the correction necessary. The effective area, which will give us the work done by the piston, is that enclosed by lines of the compression stroke, and of the working stroke. These take place during one revolution of the crankshaft. The next revolution provides the exhaust stroke, when the spent gases are pushed out of the cylinder, and the suction stroke when the new mixture is taken in ready for the whole cycle of four strokes and two revolutions to begin again. Looking at the indicator diagram you will see that there is virtually no effect on the enclosed area for the exhaust and suction strokes, and so as far as providing work done they are wasted and are neglected in the calculations.

Now $$\text{power} = \frac{\text{work done}}{\text{time taken}}$$

The power which we obtain from the diagram is called 'indicated power', but this is subject to considerable loss (10–20%) before it is useful output called 'brake power' (so named because it is usually measured using a device called a brake drum).

Task 1: to find the indicated horsepower.

You should first choose a suitable numerical rule for finding the area of the indicator diagram. The result will be in units of

$$\text{inches} \times \frac{\text{lbf}}{\text{in}^2}$$

and if you then multiply by the piston area, on which the pressure acts, in units of in^2 you will finish with units of inches lbf. Divide by 12 to obtain ft lbf units.

Now since we only have one working revolution in every two, then the work done which you have found will occur $\frac{1}{2} \times 1500$ or 750 times per minute.

Thus the indicated work done takes place in $\frac{1}{750}$ of a minute.

So, using the power equation, we can obtain the indicated work done in ft lbf/min units. Now in imperial units there is a famous unit called the horsepower (reputably so called because it was the rate at which a very strong horse could work).

Since 1 horsepower = 33 000 ft lbf/min we can now find the indicated horsepower of the cylinder—multiply by four if for the whole engine.

Task 2: to bring it all up-to-date

You should redraw the indicator diagram using SI units and rework all the calculations in SI, finishing with powers in units of watts (no horses now!).

Unit coverage

The table below shows which parts of the unit have been covered in this assignment and where there are opportunities to assess core skills. If mapping of this type is shown for all your portfolio work, it is easy to check that you have covered all the elements in the unit.

Task	Unit	Element/PC	Core skills
1	8	1.1, 1.3, 3.2, 3.3, 3.7	Application of number 3.2.1 and 3.2.6
2	8	1.1, 3.1, 3.2, 3.3	Application of number 3.3.1, 3.3.3 and 3.2.6
			Communication 3.3.2 and 3.3.3

GNVQ assessment

Here you will find the requirements and suggested procedure for the portfolio part of the unit. There then follows some useful advice on tackling multi-choice questions, together with six sample unit tests.

15 Portfolio and external test preparation

To achieve the Advanced GNVQ Mathematics for Engineering unit, you are required to successfully complete two parts.

A portfolio of work: this provides evidence that you are able to solve engineering problems in using mathematical techniques. This evidence may take many forms; for example, written assignments, laboratory experiments or records of work experience. The portfolio is assessed by your tutor against specified marking criteria and is then verified by the internal verifier (from your centre), and finally by the external verifier who will visit your study centre.

An external test: this is set by people outside your study centre. It consists of a series of multi-choice style equations which you must complete in 1 hour. You can take the test at any of the three test dates: January, April and June. The pass mark is 70%. You are allowed to retake the test, if you fail, at the next test date. There is an additional test series in September which is specifically for students who need to retake tests.

The portfolio of work

What is a portfolio?

The *portfolio* provides the evidence that the student has developed the skills, knowledge and understanding required to fulfil the role of an engineer. As such the portfolio must contain evidence of the student's work in solving engineering problems and dealing with engineering situations.

The portfolio need not just cover the Mathematics unit but may provide evidence for other GNVQ units. Evidence from one assignment may well cover the criteria for a number of units. For example, an assignment set for the Science unit may also provide the required evidence for the Mathematics unit. Because of this, it is particularly important to index your work well, and to cross reference where an assignment covers two or more units. The evidence that is required to pass the GNVQ Advanced Engineering programme is clearly stated in each of the units under the heading of *evidence indicators*. Candidates must prove that they are competent in these.

What goes into the portfolio?

You can include any work you have done which shows that you have developed your skills, knowledge and understanding of the criteria in the units. The *portfolio problems and assignment* at the end of each part of this book would be good examples of evidence for the Mathematics unit. It is important to note the following points.

The work must be *authentic*—you must have done the work yourself. Work your friend or tutor has done is not acceptable, so do not include notes and hand-outs your tutor has done for you.

The work must be *valid, relevant and complete*—you must provide evidence of work that covers every evidence indicator. The work must be relevant to the unit. If you are to be an engineer, then it is evidence of solving engineering problems which is relevant, not of any other disciplines.

The work must be *indexed*—it is your responsibility to show evidence, not the assessor's job to find it. It is therefore your job to index the portfolio to enable the assessor to find the evidence he or she is looking for in the evidence indicators.

The format, size and presentation of the portfolio

You can use any medium you wish to present your work, as long as it satisfies the evidence indicators. Some examples are given below but these are far from being exhaustive.

Graphical solutions
- a manually produced graph, chart or table
- a computer generated graph, chart or table
- a Computer Aided Drawing

Calculations
- hand written calculations
- a log book
- a work experience report by the employer
- computer output
- a sheet of rough checks/estimations
- a laboratory report

Oral presentation
- video evidence
- audio tape recording
- signed record of oral assessment by a tutor

Records of solutions
to engineering problems   – hand written assignment
– wordprocessed assignment
– science experiment
– laboratory work
– work experience reports
– log book
– photographs

The size of the portfolio is not important and will vary from student to student. It is seldom that the largest portfolio is the best or the smallest the worst—the key factor is whether the work is relevant to the evidence indicators. The portfolio often takes the form of one or two lever arch files supported by additional information such as video/audio tapes, models and manufactured components.

Presentation—this does not just mean how neat and tidy your work is, but if it has been laid out well using an appropriate medium which is relevant to the task. Are your calculations easy to follow? Could a customer understand the work you have produced or, more importantly, could the assessor? All of these points are important, but do you actually answer the problem which has been set or have you just written what you are able to do? Remember a good portfolio may not be well presented, but a poor portfolio is never well presented.

How is your portfolio graded?

Your portfolio is graded against set criteria. There are four grading themes, which your work is marked against. These are:

1) *Planning*: how you approach and monitor the tasks when working on an assignment.

2) *Information seeking and handling*: the way you identify and use information and check that it is valid.

3) *Evaluation*: the way you review and indicate possible alternative approaches to improve your assignment.

4) *Outcomes*: how well your assignment demonstrates your knowledge, skill and understanding of the subject including your command of language.

Each of these themes has two criteria, as shown in the table overleaf. You will receive grades for these criteria. The grade for each assignment may be either Pass, Merit, Distinction or no grade (fail). These grades will be collected together to give an overall grade for the process and the outcome of portfolio work. The final grade for your portfolio will be the lowest of the two overall grades for process and outcome.

Heading	Theme	Criteria
Process	Planning	1. Drawing up plans of action
		2. Monitoring courses of action
	Information seeking and handling	3. Identifying and using sources to obtain information
		4. Establishing the validity of information
	Evaluation	5. Evaluating outcomes and alternatives
		6. Justifying particular approaches to tasks/activities
Outcome	Quality of outcome	7. Synthesis
		8. Command of language

Who is going to see your portfolio?

Your tutor is the person who assesses the portfolio and the internal verifier checks that the marking is fair. It is then the role of the external verifier to check that the evidence is valid and correct. Your tutor and the verifiers have to assess whether you have sufficient work to satisfy the evidence indicators against which you are assessed. The portfolio work is yours and as such it may be useful to show to prospective employers. At the end of your time on the GNVQ course you will have gathered a large amount of work. Make sure you *do not lose it*. Your study centre may want you to store it with them for this reason.

Indexing and cross referencing

It is the responsibility of the student to ensure that the assessor can find the evidence required for the GNVQ units within the portfolio. It is not sufficient to simply say that the evidence exists in the folder.

One of the most important tasks in creating a portfolio is to index the work, so that it is possible to find particular pieces of work within the folder. This will often require you to cross reference assignments to a number of different evidence indicators. Page numbers or assignment numbers are often very useful.

A good way to check that your index works is to ask a friend to use it to find one of the evidence indicators in your portfolio, or even do this yourself.

The external test

Preparing for the external test

Make sure that you plan your revision of the course work over as long a period as possible, so that you do not run short of time as the test approaches.

You should test your understanding of the main topics covered by working some of the multi-choice style questions in Chapter 16. To achieve a pass in the test it is important to have mastered the basic ideas of the unit. Do not waste revision time on material that is not covered in the test. Check with your tutor which topics are actually tested; this information is available in the unit test specification. It may be helpful to ask the advice of your tutor on producing a revision plan. He or she should know your strengths and weaknesses and therefore on which topics you should concentrate your revision. Remember it is the quality of revision you do that is important, not the amount of time you spend.

A specimen/mock external test paper

Before the actual test it is a good idea to have a go at a specimen/mock test paper. Your tutor should let you have a copy of one. This will enable you to know exactly what to expect, particularly the time constraints. Look carefully at the results of this paper and schedule any further revision based on this.

The external test paper

The test is 1 hour long and will consist of 22 multi-choice style questions. (Check this with your tutor, from the test specification.) Each question carries one mark. To achieve a pass grade you need to correctly answer 15 or more questions.

Technique for answering questions

The best technique when answering questions is to be familiar with and fully understand the subject. However, here are some points which will hopefully improve your score:

1) Always read the question carefully.
2) Read *all* the answers before you make your selection. Do not stop reading the remaining answers because you think A, B, or C is the answer. It may be very similar to the answer, but not the correct answer.
3) Use reverse logic for questions where you cannot find the right answer. Eliminate the options which you know are not correct. Then make a calculated guess at the remaining options. You have a 50–50 chance of getting the answer correct if you can eliminate two options.

4) Always select an answer, even if you have to guess. One of the four options is correct and you may be lucky. If you do not select an answer you can never get it right.

5) It will help to use the shaded boxes for working out the correct answers.

Advice for the test day

Be fully prepared—make sure that you have your calculator, an HB pencil and an eraser, and that you arrive at the examination centre in plenty of time, without being rushed. Allow extra time in case of any mishaps or delays.

Read the paper carefully—this sounds very obvious but under the pressure of a test it is easy to misinterpret the true or false statements, particularly if your friends surrounding you have started at a faster pace than yourself.

Do not panic—this is easy to say, but you need to stay calm throughout the test. If after about 3 minutes on one question you are no closer to finding the answer, or you are getting bogged down with a question, then start a different question. It often happens that when you come back to a question you have a better idea, or spot a mistake you had missed previously.

State the answer clearly—a computer will be marking your answer paper. It can only accept one answer. When you have decided which option correctly represents the answer to a question, carefully mark this on the paper opposite the appropriate question number. It is difficult and time consuming to move all your answers down a line if you have missed out a question.

Use your time effectively—if you are running short of time towards the end of the test, tackle first the questions that look straightforward, and so on. Do not get bogged down, but make sure you have an answer for every question. Even if you have to guess, there is a 1 in 4 chance you will guess correctly and no marks are deducted for incorrect answers.

Check your answers—a surprisingly large number of students are careless when completing their answer sheet. Make sure you fill in the details at the top of the page and sign your name at the bottom. Check that each of the answers you have chosen is against its correct question number.

Good luck—but if you have prepared well, you will not need it...

16 Sample unit tests

Sample unit test 1

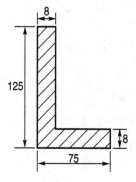

Fig. ST1.1

1) The binary equivalent of the denary number 93 is

   A 1011101   **B** 1101101   **C** 111101   **D** 10101101

2) If 1 inch = 25.4 mm and 1 foot = 12 inches, how many square feet are there in 1 square metre, correct to 3 s.f.?

   A 10.8   **B** 18.1   **C** 108   **D** 129

3) A round steel bar 80 mm diameter and 500 mm long is made from material having a density of 7800 kg/m³. Correct to 3 s.f., what is the mass of the bar in kilograms?

   A 1.96   **B** 19.6   **C** 25.6   **D** 196

4) An unequal angle cross-section of a 2 m long steel tie bar is as shown in Fig. ST1.1. If the density of steel is 7800 kg/m³, find the mass of the tie bar in kilograms, correct to 3 s.f.

   A 22.0   **B** 23.0   **C** 24.0   **D** 25.0

5) The law of expansion of a gas is given by $pV^{1.3} = k$. If pressure $p = 0.64 \times 10^6$ and the volume $V = 0.31$, find the value of k correct to 3 s.f.

   A 139×10^3 **B** 1.39×10^6 **C** 140×10^3 **D** 1.40×10^6

6) The number 0.063 92 may be stated as:

   (i) 0.064 correct to 3 d.p.
   (ii) 0.064 correct to 3 s.f.

   Decide whether each of these statements is true or false.

   A (i) is true and (ii) is true   **B** (i) is true and (ii) is false
   C (i) is false and (ii) is true   **D** (i) is false and (ii) is false

7) If 0.1% is the relative percentage accuracy of a 0.5 km measurement then its maximum absolute error is

   A 0.05 m   **B** 0.5 m   **C** 5 m   **D** 50 m

8) We may write $\log \dfrac{ab}{c^3}$ as

   A $\dfrac{\log ab}{3 \log c}$ **B** $\dfrac{\log a + \log b}{3 \log c}$

   C $\log ab - (\log c)^3$ **D** $\log a + \log b - 3 \log c$

9) When simplified $\quad 3a + 2a \times 6a - 4a \div 2 \quad$ becomes

    A $\quad 12a^2 + a$ **B** $\quad 30a^2 - 2a$ **C** $\quad 20a^2$ **D** $\quad 5a^2$

10) Removing bracket from $\quad (x^2 + 1)(1 - x) \quad$ gives

    A $\quad x^2 - 2x + 1$ **B** $\quad 2x^2 - 2x + 1$ **C** $\quad 1 - x^3$
    D $\quad -x^3 + x^2 - x + 1$

11) The expression $\quad 2pr - 4ps + qr - 2qs \quad$ when factorised becomes

    A $\quad (2p + q)(r - 2s)$ **B** $\quad (2p + q)(2s - r)$
    C $\quad (2ps - q)(r + p)$ **D** $\quad (2p + s)(q - r)$

12) The resistance R of a wire after a temperature rise t is given by $R = R_0(1 + \alpha t)$ where R_0 is zero temperature resistance and α is a coefficient. Transposing the equation for t gives

    A $\dfrac{R - R_0}{\alpha}$ **B** $\dfrac{R}{R_0} - \alpha$ **C** $\dfrac{1}{\alpha}\left(1 - \dfrac{R}{R_0}\right)$ **D** $\dfrac{R - R_0}{R_0 \alpha}$

13) The value of x correct to 3 s.f. if $\quad 3x + 6 = 19 - 2x \quad$ is

    A $\quad -20.5$ **B** $\quad 0.571$ **C** $\quad 2.60$ **D** $\quad 17.1$

14) The solutions of the equation $\quad x^2 - 3x = 0 \quad$ are

    A $\quad 0, 0$ **B** $\quad 0, -3$ **C** $\quad 0, 3$ **D** $\quad$ No solution

15) Which sketch shows a quadratic equation with roots $x = 1$ and $x = -2$?

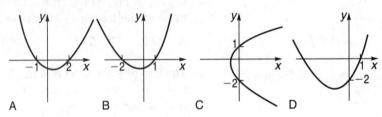

A   B   C   D

16) Eliminating y from the equations $3x - 4y = -10$ and $x + 4y = 8$ we get

    A $\quad 2x = -18$ **B** $\quad 4x = -18$ **C** $\quad 2x = -2$ **D** $\quad 4x = -2$

17) A washer has outer and inner diameters of D and d, respectively, and its area is given by $\frac{\pi}{4}(D^2 - d^2)$. If $D = 21.4\,\text{mm}$ and $d = 15.9\,\text{mm}$ then the value of the area, correct to 3 s.f., is

    A $\quad 4.32\ \text{mm}^2$ **B** $\quad 40.3\ \text{mm}^2$ **C** $\quad 161\ \text{mm}^2$ **D** $\quad 558\ \text{mm}^2$

18) A steel plate is shown in Fig. ST1.2. The distance x, to the greatest accuracy the given dimensions allow, is

    A $\quad 78.8\ \text{mm}$ **B** $\quad 85.3\ \text{mm}$ **C** $\quad 286\ \text{mm}$ **D** $\quad 297\ \text{mm}$

19) The plan of a plastic plate with two holes drilled in it is shown in Fig. ST1.3. Correct to 1 d.p., the coordinate dimensions a and b are

    A $\quad 46.8, 41.5$ **B** $\quad 48.8, 38.5$ **C** $\quad 55.8, 44.5$ **D** $\quad 56.8, 39.5$

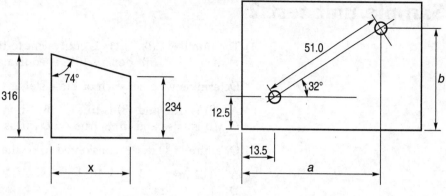

Fig. ST1.2   Fig. ST1.3

20) Two voltage phasors are shown in Fig. ST1.4. The resultant voltage and the angle it makes with V_1, giving both answers correct to 3 s.f., are

   A   114 V, 26.9°   **B**   114 V, 29.6°   **C**   141 V, 26.9°
   D   141 V, 29.6°

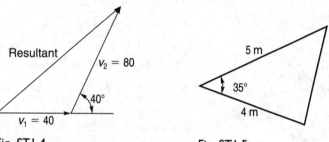

Fig. ST1.4   Fig. ST1.5

21) In square metres, correct to 3 s.f., the area of the triangle shown in Fig. ST1.5 is

   A   5.74   **B**   7.54   **C**   10.0   **D**   20.0

22) Correct to 3 s.f., the dimensions x and y in Fig. ST1.6 are

   A   8.49, 8.60   **B**   8.52, 8.60   **C**   8.57, 8.60   **D**   8.60, 8.49

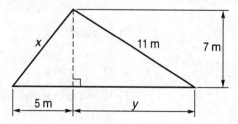

Fig. ST1.6

Sample unit test 2

1) The number   (i) 0.11_2 is equivalent to the number 0.75_{10}
(ii)   denary 1011 is equivalent to binary number 11.

Determine whether each of these statement is true or false.

A   (i) is true and (ii) is true   B   (i) is true and (ii) is false
C   (i) is false and (ii) is true   D   (i) is false and (ii) is false

2) One litre of water is considered to be the same as

A 0.01 m^3 B $1 \times 10^{-3} \text{ m}^3$ C $1 \times 10^{-6} \text{ m}^3$
D $1 \times 10^{-9} \text{ m}^3$

3) A length of 50 mm measured to the nearest millimetre has a maximum absolute error of

A -0.5 mm   B   0.5 mm   C   1 mm   D   1%

4) A temperature of 23.6°C is measured correct to the nearest $\frac{1}{10}$°C. the relative percentage accuracy correct to 3 s.f. is

A   0.212%   B   0.424%   C   2.12%   D   4.24%

5) An aluminium tube is as shown in Fig. ST2.1. In kilograms the mass of the tube, correct to 3 s.f. if aluminium has density 2700 kg/m^{-3}, is

A   17.7   B   19.8   C   21.0   D   22.4

6) The total (inner and outer) surface area of the tube shown in Fig. ST2.1, correct to 3 s.f. in units m^2, is

A   0.758   B   7.58   C   75.8   D   758

7) Evaluating, correct to 3 s.f., the expression $\dfrac{2 \times 10^5 + 3 \times 10^4}{6 \times 10^6}$ gives

A   0.0383   B   0.383   C   83.3   D   833

8) If $y = ax^b$ then we have for $\log y$

A $\log a + \log x + \log b$ B $\log a + \log x - \log b$
C $\log a + b \log x$ D $\log a + \log bx$

9) An alternative to $1 - \dfrac{x}{x-2}$ is

A $\dfrac{2}{x-2}$ B $\dfrac{2x-2}{x-2}$ C $\dfrac{2}{2-x}$ D $\dfrac{1-x}{x-2}$

10) Expand $(a - 2b)^2$ and obtain

A $a^2 - 2b + b^2$ B $a^2 - 4ab - 4b^2$ C $a^2 - 4ab + 4b^2$
D $a^2 + 4ab + 4b^2$

11) When factorised $a^2 - 2ab + b^2$ becomes

A $(a+b)(a-b)$ B $(a^2-1)(1-b^2)$ C $(a+b)^2$
D $(a-b)^2$

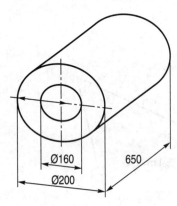

Ø160   650
Ø200

Fig. ST2.1

12) Temperature in degrees fahrenheit F and in degrees celcius C is related by $F = \dfrac{9}{5}C + 32$. Transposing gives an expression for C

   A $32 - \dfrac{5}{9}F$ **B** $\dfrac{5(F - 32)}{9}$ **C** $\dfrac{9}{5}F - 32$ **D** $\dfrac{5}{9}F - 32$

13) If $5.6(20 - 6.3m) = m$ then the value of m, correct to 3 s.f., is

   A   1.68   **B**   1.78   **C**   3.09   **D**   10.7

14) Using the quadratic formula to solve $x^2 + 3x + 1 = 0$ correct to 2 d.p. gives

   A $-0.38, -2.62$ **B** $-0.38, 2.62$ **C** $0.38, -2.62$
   D $0.38, 2.62$

15) The sketch which shows the quadratic equation $x^2 - 4 = 0$ and its roots is

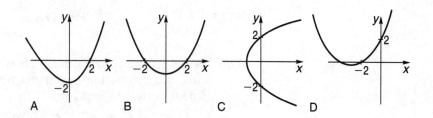

   A   B   C   D

16) Two numbers, x and y, are such that their sum is 18 and their difference is 12. The values of x and y may be found from simultaneous equations

   A $\begin{aligned} x - y &= 18 \\ y - x &= 12 \end{aligned}$ **B** $\begin{aligned} x + y &= 18 \\ y - x &= 12 \end{aligned}$ **C** $\begin{aligned} x - y &= 18 \\ x + y &= 12 \end{aligned}$

   D $\begin{aligned} y - x &= 18 \\ y + x &= 12 \end{aligned}$

17) A hydraulics formula for volume of flow is $Q = 2.37H^{5/2}$. If the head $H = 2.81$ then the value of Q, correct to 3 s.f., is

   A   18.7   **B**   19.9   **C**   31.4   **D**   114

18) In Fig. ST2.2 the distance d, to the accuracy of the numbers given, is

   A   12.1 mm   **B**   19.0 mm   **C**   20.9 mm   **D**   33.0 mm

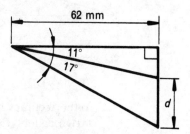

Fig. ST2.2

19) A plate has four holes drilled in it on a pitch circle of 80 mm diameter as shown in Fig. ST2.3. Correct to $\frac{1}{10}$ mm, the rectangular coordinates of hole C are

   A (−36.4, −20.0)   **B** (−34.6, 20.0)   **C** (34.6, −20.0)
   D (34.6, 20.0)

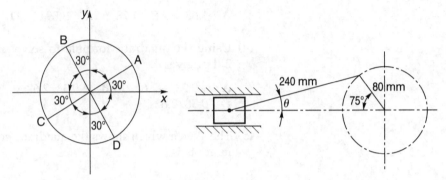

Fig. ST2.3   Fig. ST2.4

20) The 'slider-crank'arrangement shown in Fig. ST2.4 shows piston, crank and connecting (conn) rod for a reciprocating engine. The angle θ, correct to 3 s.f., is

   A 15.0°   **B** 18.8°   **C** 19.7°   **D** 71.2°

21) ABCD is a parallelogram which represents the shape of a lamina template, shown in Fig. ST2.5. Its area, correct to the nearest square millimetre, is

   A 2479 mm^2 **B** 2749 mm^2 **C** 5497 mm^2
   D 5974 mm^2

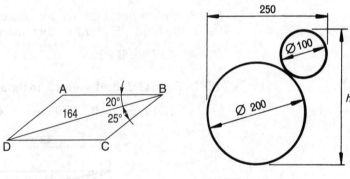

Fig. ST2.5   Fig. ST2.6

22) To the accuracy of the given data, the distance h in the arrangement of two rollers, Fig. ST2.6, used for a metrology measurement, is

   A 250 mm   **B** 301 mm   **C** 262 mm   **D** 412 mm

Sample unit test 3

1) The denary equivalent of the binary number 11.01 is

    A 2.35   **B** 2.53   **C** 3.25   **D** 3.52

2) (i) A micron (10^{-6} m) is a preferred unit
   (ii) A centimetre is a preferred unit

Decide whether each of these statments is true or false.

    A (i) is true and (ii) is true   **B** (i) is true and (ii) is false
    C (i) is false and (ii) is true   **D** (i) is false and (ii) is false

3) A tiny mass of 0.496 mg measured correct to the nearest μg has a maximum value of

    A 496.5 μg   **B** 0.496 mg   **C** 497 μg   **D** 0.497 mg

4) The mass of a casting is 539 kg correct to the nearest kg. Its relative percentage accuracy, correct to 3 d.p., is

    A 0.0928%   **B** 0.093%   **C** 0.186%   **D** 0.928%

5) A universal beam has dimensions as shown in Fig. ST3.1. If the density of steel from which it is made is 7800 kg m^{-3}, correct to 3 s.f., the mass per metre unit length of the beam is

    A 67 kg m^{-1} **B** 176 kg m^{-1} **C** 206 kg m^{-1}
    D 319 kg m^{-1}

6) A cylindrical storage container is 900 mm in diameter and 2 metres long overall, and has hemispherical ends. Its capacity to the nearest litre is

    A 1018 ℓ **B** 1039 ℓ **C** 1066 ℓ **D** 1081 ℓ

7) A formula used to find the stiffness of a close coiled helical spring is $\dfrac{Cd^4}{8D^3n}$ and if $d = 5.6$, $D = 53$, $n = 10$ and $C = 83 \times 10^3$ the value of the stiffness, correct to 3 s.f., is

    A 5.68   **B** 5.86   **C** 6.58   **D** 6.85

8) If $1.3^x = 3^{0.7}$ then x is given by the expression

    A $\dfrac{0.7(\log 3)}{\log 1.3}$ **B** $\dfrac{3(\log 0.7)}{\log 1.3}$ **C** $0.7 \log 3 - \log 1.3$

    D $\dfrac{\log(0.7 \times 3)}{\log 1.3}$

9) $\dfrac{1}{mn} + \dfrac{1}{m} + 1$ is the same as

    A $\dfrac{3}{mn}$ **B** $\dfrac{1+m+n}{mn}$ **C** $\dfrac{1+m+mn}{mn}$ **D** $\dfrac{1+n+mn}{mn}$

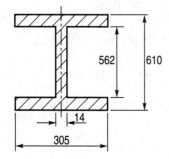

562 |610

14

305

Fig. ST3.1

10) Multiplying out $(x+3)(2x-2)$ gives

   A $2x^2+4x-4$ **B** $2x^2+4x+6$ **C** $2x^2+4x-6$
   D $2x^2+8x+6$

11) Factorising $b(x+y)-c(y+x)$ gives

   A $(b-c)(x-y)$ **B** $(b-y)(x-c)$ **C** $(b-x)(c-y)$
   D $(b-c)(x+y)$

12) The percentage profit P made in a transaction is given by $P = \dfrac{s-b}{b} \times 100$ where s is the selling price, and b the buying price. Transposing for b gives

   A $s+\dfrac{Ps}{100}$ **B** $s+1-\dfrac{P}{100}$ **C** $\dfrac{s}{100P+1}$

   D $\dfrac{100s}{P+100}$

13) Given that $\dfrac{m+1}{2} = \dfrac{1-m}{3}$ then, correct to 3 d.p., the value of m is

   A -0.964 **B** -0.200 **C** 0.200 **D** 0.964

14) By finding the factors of, or by using the formula, the roots of the equation $x^2-x+2=0$ are

   A $-2,-1$ **B** $2,-1$ **C** $-2,1$ **D** No solution

15) The sketch representing the equation $(x-1)^2$ and its roots is

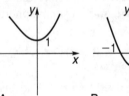

   A   B   C   D

16) Eliminating x from the equations $\begin{aligned} 3x+5y &= 2 \\ x+3y &= 7 \end{aligned}$ we get

   A $4y=-19$ **B** $8y=9$ **C** $4y=19$ **D** $4y=5$

17) The shear stress τ in the torsion of a circular shaft is given by $\tau = \dfrac{Tr}{J}$. If the radius $r=25$, torque $T=900\times10^3$ and the second moment of area $J=6.13\times10^5$, then the shear stress, correct to 3 s.f., is

   A 0.367 **B** 3.67 **C** 36.7 **D** 367

18) A groove for a vee belt is shown in Fig. ST3.2. Then to the accuracy appropriate to the given data the distance W is

   A 2.43 mm   **B** 3.76 mm   **C** 4.29 mm   **D** 4.37 mm

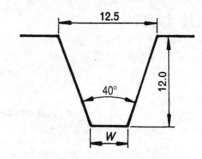

Fig. ST3.2

19) Two points P and Q have coordinates as shown in Fig. ST3.3. Then, correct to 3 s.f., the polar coordinates of the mid-point of PQ are

    A (23.3, 36.9°)   **B** (24.5, 40°)   **C** (25.4, 42.0°)
    D (26.0, 40.0°)

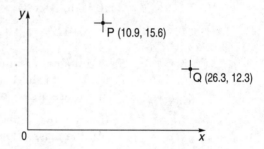

Fig. ST3.3

20) The distance PQ in Fig. ST3.3, correct to 3 s.f., is

    A 14.9 mm   **B** 15.2 mm   **C** 15.4 mm   **D** 15.7 mm

21) One version of the cosine rule is $b^2 = a^2 + c^2 - 2ac \cos B$. Transposed for the angle B this formula becomes

    A $\text{inv} \cos \dfrac{a^2 + c^2}{a^2 + 2ac}$ **B** $\dfrac{a^2 + c^2 - b^2}{2ac}$

    C $\text{inv} \cos \dfrac{a^2 + c^2 - 2ac}{b^2}$ **D** $\text{inv} \cos \dfrac{a^2 + c^2 - b^2}{2ac}$

22) The area of the triangle shown in Fig. ST3.4, correct to 3 s.f., is

    A 173 mm²   **B** 223 mm²   **C** 273 mm²   **D** 332 mm²

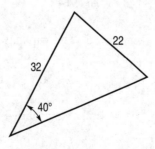

Fig. ST3.4

Sample unit test 4

1) The binary sum of 1110_2 and 1101_2 is

   A 10111   **B** 10010   **C** 11001   **D** 11011

2) A car accelerates at 1 m/s^{-2} which is equivalent, in units km h^{-2}, to

   A 3.6   **B** 12.96   **C** 3600   **D** 12 960

3) The voltage 0.060 V measured to the nearest millivolt (mV) is

   A 0.060 ± 0.0001 V   **B** 0.060 ± 0.0005 V
   C 0.060 ± 0.001 V   **D** 0.060 ± 0.005 V

4) The relative percentage accuracy of 7850 W correct to the nearest 10 W to 3 s.f. is

   A 0.0127%   **B** 0.0637%   **C** 0.127%   **D** 0.637%

5) A container is symmetrically shaped as shown in Fig. ST4.1, and is made from sheet steel which has a mass of 18 kg per square metre of surface area. Correct to 3 s.f.,

   (i) the total outer surface area is 0.911 m^2
   (ii) the mass of the container is 14.6 kg

   Decide whether each of these statements is true or false.

   A (i) is true and (ii) is true   **B** (i) is true and (ii) is false
   C (i) is false and (ii) is true   **D** (i) is false and (ii) is false

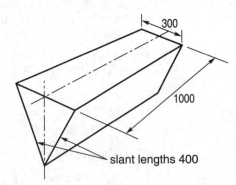

Fig. ST4.1

6) The capacity of the container shown in Fig. ST4.1, correct to the nearest litre, is

   A 56 ℓ **B** 59 ℓ **C** 60 ℓ **D** 61 ℓ

7) A simplified form of $\sqrt[5]{x^{-2}}$ is

   A x^3 **B** x^{-10} **C** $\dfrac{1}{x^{5/2}}$ **D** $\dfrac{1}{x^{2/5}}$

8) If $8.3^x = 4.9$ then x is given by the expression

  A $\log 4.9 - \log 8.3$ **B** $\log 8.3 - \log 4.9$

  C $\dfrac{\log 4.9}{\log 8.3}$ **D** $\log\left(\dfrac{4.9}{8.3}\right)$

9) The expression $\dfrac{2}{x} + \dfrac{3}{x-1}$ may be written as

  A $\dfrac{5x-2}{x^2-1}$ **B** $\dfrac{5x-3}{x(x-1)}$ **C** $\dfrac{5x-2}{x(x-1)}$ **D** $\dfrac{5x}{x^2-1}$

10) Expanding $\dfrac{x}{2}(x-1)(x+1)$ gives

  A $\dfrac{x^3}{2} - \dfrac{x}{2}$ **B** $\dfrac{x^3}{2} - x^2 - \dfrac{x}{2}$ **C** $\dfrac{x^3}{2} + \dfrac{x^2}{4} - x + 1$

  D $x^3 + \dfrac{x^2}{2} - x - 4$

11) The expression $4m^2 + 12mn + 9n^2$ has factors

  A $(4m+9n)(m+n)$ **B** $(4m+n)(m+9n)$
  C $(4m^2+1)(1+9n^2)$ **D** $(2m+3n)^2$

12) Current I and voltage V in a circuit with two resistances R and r in series are related by $I = \dfrac{V}{R+r}$. An expression for R is

  A $\dfrac{I}{V} - r$ **B** $\dfrac{V}{I} - r$ **C** $\dfrac{1}{I}(V-r)$ **D** $\dfrac{1}{V}(V-Ir)$

13) If $8.11 + \dfrac{1}{t} = \dfrac{6}{t}$ then the value, correct to 3 s.f., of t is

  A -1.62 **B** 0.617 **C** 1.62 **D** 3.11

14) Use of factors to solve $3x^2 - 2x - 5 = 0$ gives

  A $-\dfrac{5}{3}, -1$ **B** $-\dfrac{5}{3}, 1$ **C** $\dfrac{5}{3}, -1$ **D** $\dfrac{5}{3}, 1$

15) The equation represented by the graph sketched in Fig. ST4.2 is

  A $(x+3)(x-1) = 0$ **B** $(x-3)(x+1) = 0$
  C $(x-3)(x-1) = 0$ **D** $(x+3)(x+1) = 0$

16) The equations $\begin{array}{l} 3x - 2y = 5 \\ 4x - y = 10 \end{array}$ have solutions

  A $x = -3, y = 22$ **B** $x = -3, y = -22$
  C $x = 3, y = 2$ **D** $x = 3, y = -2$

17) The period t seconds of a simple pendulum is given by $t = 2\pi\sqrt{\dfrac{l}{g}}$ where l metres is the length and g m s^{-2} the gravitational constant. The value of t, to 2 d.p., if $l = 1370$ mm and $g = 9.81$ m s^{-2} is

  A 0.88 s   **B** 1.07 s   **C** 1.53 s   **D** 2.35 s

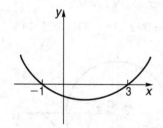

Fig. ST4.2

18) In Fig. ST4.3

 (i) the difference in length between AB and DB is 6.35 mm
 (ii) the angle ABD is 32.5°

Decide whether each of these statements is true or false.

A (i) is true and (ii) is true
B (i) is true and (ii) is false
C (i) is false and (ii) is true
D (i) is false and (ii) is false

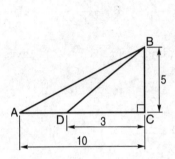

Fig. ST4.3

Fig. ST4.4

19) In the circular plate shown in Fig. ST4.4 a sector of 54° is cut out and the remainder divided to give eight equally spaced holes on a pitch circle of diameter 60 mm. Relative to Cartesian axes, which intersect at the inner point of the sector vee, the rectangular coordinates of hole P, correct to 3 s.f., are

A (−8.77, −28.7)
B (8.77, 28.7)
C (8.77, −28.7)
D (−8.77, 28.7)

20) In a state of equilibrium, mass $m = 5$ kg is on an incline as shown in Fig. ST4.5, where $\theta = 25°$. The values of forces F and N in that order, both in newtons and correct to 3 s.f., are

A 20.7, 44.5
B 22.9, 54.1
C 53.3, 24.9
D 55.5, 29.4

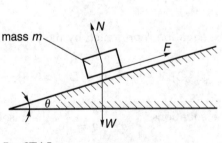

Fig. ST4.5

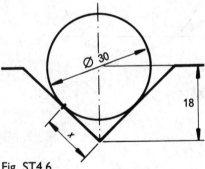

Fig. ST4.6

21) In the arrangement shown in Fig. ST4.6 the dimension x, correct to 3 s.f., is

A 6.95 mm
B 7.57 mm
C 8.95 mm
D 9.95 mm

22) The area enclosed between the bottom of the roller and the sides of length x of the vee in Fig. ST4.6, correct to 3 s.f., is

A 15.7 mm²
B 17.5 mm²
C 19.3 mm²
D 21.2 mm²

Sample unit test 5

1) The bicimal 0.0101 has a decimal equivalent of

   A 0.3125 **B** 0.3215 **C** 0.3251 **D** 0.3521

2) Given that 1 mile = 1.6 kilometres, then 30 miles per hour equals

   (i) 18.75 km/h (ii) 13.3 m/s

   Decide whether each of these statements is true or false.

   A (i) is true and (ii) is true **B** (i) is true and (ii) is false
   C (i) is false and (ii) is true **D** (i) is false and (ii) is false

3) Two resistances are measured as 3.2 ohm and 7.8 ohm correct to the nearest $\frac{1}{10}$ ohm. When connected in series the effective resistance is their sum which, in ohms, may be given as

   A 11.0 **B** 11.0 ± 0.05 **C** 11.0 ± 0.1 **D** 11.0 ± 0.5

4) If a measurement has an absolute accuracy of 0.5 mm, and a relative percentage error of 1.064, then the measurement, correct to 3 s.f., is

   A 5.32 mm **B** 21.3 mm **C** 47.0 mm **D** 532 mm

5) A 2 metre length of copper bar has a hexagonal cross-section, the length of each flat being 30 mm. If the density of copper is 8800 kg m^{-3} correct to 3 s.f. then the mass of the bar is

   A 18.8 kg **B** 21.9 kg **C** 22.4 kg **D** 41.2 kg

6) It is required to replace two pipes with bores of 32 mm and 60 mm respectively with a single pipe which has the same area of flow. The bore of the single pipe, to the nearest millimetre, is

   A 66 mm **B** 67 mm **C** 68 mm **D** 69 mm

7) Writing $\dfrac{(\sqrt[3]{u})(u)}{u^2}$ in a simplified form gives

   A $u^{-1/3}$ **B** $u^{-2/3}$ **C** u^2 **D** $u^{1/3}$

8) If $1.3^{x+2} = 7^x$ then x is given by

   A $\dfrac{2 \log 1.3}{\log 7 - \log 1.3}$ **B** $\dfrac{2 \log 1.3}{\log 1.3 - \log 7}$

   C $\dfrac{\log 7}{\log 1.3} - 2$ **D** $\dfrac{2 \log 7}{\log 1.3}$

9) We may write $\dfrac{1}{a} + b - \dfrac{c}{d}$ as

   A $1 - abd - ac$ **B** $\dfrac{1 + b - c}{ad}$

   C $\dfrac{d + abd - ac}{ad}$ **D** $\dfrac{1 + bd - c}{ad}$

10) Removing brackets from $x(x-1) + \dfrac{1}{x}(x^2+1)$ gives

   A x^2+1 **B** x^2+x **C** $2x^2+x+\dfrac{1}{x}$ **D** $x^2+\dfrac{1}{x}$

11) Factorising $3p^2+p-2$ gives

   A $(p-1)(3p-2)$ **B** $(3p+1)(p-2)$ **C** $(p+1)(3p-2)$
   D $(3p+1)(3p-2)$

12) A sphere of radius R has a cap, of height h, cut off its top. The volume V of the cap is given by $V = \dfrac{\pi h^2}{3}(3R-h)$. An expression for R is

   A $\dfrac{\pi h^3}{3V}-1$ **B** $\dfrac{3V}{\pi h}-\dfrac{h}{3}$ **C** $\dfrac{V}{\pi h^2}-\dfrac{h}{3}$ **D** $\dfrac{3V+\pi h^3}{3\pi h^2}$

13) Correct to 3 s.f., the value of x from equation $\dfrac{x-1.74}{x+0.894} = -2.96$ is

   A -0.229 **B** 0.229 **C** 2.29 **D** 22.9

14) Using the quadratic formula, correct to 3 d.p., the roots of $x = \dfrac{5x-1}{2x}$ are

   A $-0.261, -0.547$ **B** $-0.261, 0.547$ **C** $0.219, 2.281$
   D $0.291, -2.281$

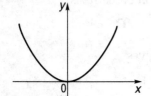

Fig. ST5.1

15) The sketch shown in Fig. ST5.1 represents a quadratic equation with

   (i)   no solutions
   (ii) a zero, repeated root

Decide whether each of these statements is true or false.

   A (i) is true and (ii) is true   **B** (i) is true and (ii) is false
   C (i) is false and (ii) is true   **D** (i) is false and (ii) is false.

16) The solutions to the simultaneous equations $2x-5y=3$ and $x-3y=1$ are

   A $x=4, y=1$ **B** $x=1, y=4$ **C** $x=13, y=4$
   D $x=4, y=3$

17) The length of a conical nozzle is 600 mm. The diameters at the ends are 150 mm and 220 mm, respectively. If the volume $V = \dfrac{1}{3}l(a + \sqrt{aA} + A)$, then the volume in litres, to the nearest $\dfrac{1}{10}$ of a litre, is

   A 8.7 **B** 16.3 **C** 17.3 **D** 18.3

18) A triangular template is shown in Fig. ST5.2. Length AB, to 3 s.f., is

A 77.7 mm   **B** 79.3 mm   **C** 82.7 mm   **D** 90.0 mm

19) The distance between the two points A and B in Fig. ST5.3, to 3 s.f., is

A 10.5   **B** 11.5   **C** 12.3   **D** 12.9

20) The area OAB in Fig. ST5.3, correct to 3 s.f., is

A 24.5   **B** 25.4   **C** 27.4   **D** 42.5

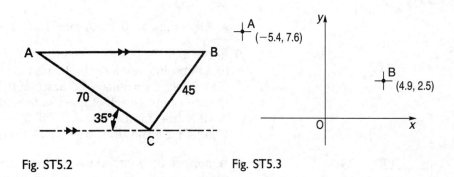

Fig. ST5.2   Fig. ST5.3

21) The holes drilled in a component are positioned by the angle and dimensions shown in Fig. ST5.4. The distance between the hole centres A and B correct to four significant figures is

A 154.5 mm   **B** 160.9 mm   **C** 162.8 mm   **D** 168.1 mm

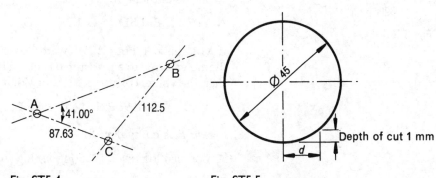

Fig. ST5.4   Fig. ST5.5

22) The set-up in Fig. ST5.5 shows the conditions which apply when a milling cutter approaches the work. Then, correct to 3 s.f., the approach distance d in millimetres is

A 6.63   **B** 7.57   **C** 8.43   **D** 9.43

Sample unit test 6

1) The number 7.75_{10} is the same as

   A 11.111_2 **B** 111.011_2 **C** 111.11_2 **D** 1101.11_2

2) A car averages 45 miles per gallon of petrol. If 1 gallon = 4.55 litres, and 1 mile = 1.61 kilometres , its equivalent in kilometres per litre, correct to 3 s.f., is

   A 15.9   **B** 30.9   **C** 44.9   **D** 116

3) Careless recording gave the dimensions of a rectangle as 50 mm long by 60 mm wide, when both measurements had, in fact, been measured to an accuracy of $\frac{1}{10}$ mm. The minimum value of the perimeter is

   A 218.0 mm   **B** 219.0 mm   **C** 219.8 mm   **D** 219.9 mm

4) In the data for question 3) the value of the perimeter has
   (i) an absolute accuracy of 0.2 mm
   (ii) a relative percentage accuracy of 0.091%, correct to 3 d.p.
   Decide whether each of these statements is true or false.
   A (i) is true and (ii) is true   **B** (i) is true and (ii) is false
   C (i) is false and (ii) is true   **D** (ii) is false and (ii) is false

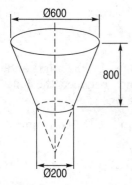

Ø600

800

Ø200

Fig. ST6.1

5) A hopper of circular cross-section, used for storing and feeding plastic pellets into an extrusion machine, is shown in Fig. ST6.1. If the volume of a cone is given by $V = \frac{1}{3}Ah$ then its capacity, to the nearest litre, is

   A 11 ℓ **B** 21 ℓ **C** 51 ℓ **D** 109 ℓ

6) If the hopper shown in Fig. ST6.1 has an open top but closed bottom, then the total (inner and outer) surface area in square metres, correct to 3 s.f., if the outer curved surface area of a cone = $\pi r \times$ slant length, is

   A 1.07   **B** 1.51   **C** 1.98   **D** 2.14

7) If volume per hour V of water passing through an injector of diameter d due to a pressure p is given by the formula $V = 2d^2p^{1/8}$ then the value, correct to 3 s.f., of V if $d = 15$ and $p = 1.5 \times 10^5$, is

   A 1990   **B** 1996.291   **C** 1996   **D** 2000

8) $\log m^2 n^3$ is equivalent to

   A $(\log m^2)(\log n^3)$ **B** $6(\log m)(\log n)$ **C** $(\log m)^2 + (\log n)^3$

   D $2 \log m + 3 \log n$

9) The expression $\dfrac{\dfrac{1}{x} + 1}{\dfrac{1}{x}}$ may be simplified to

   A $1 + \dfrac{1}{x}$ **B** $1 + \dfrac{1}{x^2}$ **C** $\dfrac{1+x}{x}$ **D** $1 + x$

10) Multiplying out $(x+1)(x-1)^2$ gives

    A $x^2 + 3x - 1$ **B** $x^3 - 1$ **C** $x^3 + x^2 - 1$ **D** $x^3 - x^2 - x + 1$

11) The factors of $\left(\dfrac{1}{x^2} - b^2\right)$ are

    A $\left(\dfrac{1}{x^2} - 1\right)(1 + b^2)$ **B** $\left(\dfrac{1}{x} - b\right)\left(\dfrac{1}{x} + b\right)$ **C** $\left(\dfrac{1}{x} - b\right)^2$

    D $(1 - b^2)\left(\dfrac{1}{x^2} + 1\right)$

12) Transposing for H from a quadrilateral area formula

$$A = \frac{a(H + h) + bh + cH}{2} \text{ gives}$$

    A $\dfrac{2A(b + c)}{ac}$ **B** $\dfrac{2A - (a + c)h}{a + c}$ **C** $\dfrac{2A - (a + b)h}{a + c}$

    D $\dfrac{2A}{(a + b)h} - (a + c)$

13) If $\dfrac{2}{x} + \dfrac{1}{3x} = 0.631$ the value of x, correct top 3 s.f., is

    A -1.06 **B** 1.06 **C** 1.70 **D** 3.70

14) Using the formula then the roots of $x^2 + 4.61x = 3.16$, correct to 3 d.p., are

    A $-0.606, -5.216$ **B** $-0.606, 5.216$ **C** $0.606, -5.216$

    D $0.606, 5.216$

15) The sketch representing the equation $x^2 + 1 = 0$ is

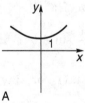   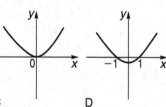

    A   B   C   D

16) A motorist travels x km at 50 km/h, and y km at 60 km/h. The total time taken is 5 hours. If the average speed is 56 km/h then

    A $50x + 60y = 5$ **B** $6x + 5y = 1500$
            $x + y = 280$ $x + y = 280$

    C $\dfrac{x}{50} + \dfrac{y}{60} = 5$ **D** $50x + 60y = 5$
                             $x + y = 280$

       $\dfrac{x + y}{5} = 56$

17) Constant acceleration a m s^{-2} for time t s causes velocity u m s^{-1} to increase to v m s^{-1} according to the equation of motion $v = u + at$. If $u = 30$ km/h and $a = 0.5$ m s^{-2} then the velocity, correct to 3 s.f., after 5 seconds is

    A 9.03 m s^{-1} **B** 10.8 m s^{-1} **C** 32.5 m s^{-1} **D** 42.0 m s^{-1}

18) The angle θ in the turned bar shown in Fig. ST6.2, correct to 3 s.f., is

    A 55.9°   **B** 73.0°   **C** 79.5°   **D** 95.3°

19) If OP $= 40$ mm and PQ $= 50$ mm then the Cartesian coordinates of Q, correct to 3 s.f., in the arrangement shown in Fig. ST6.3 are

    A 30.6, 25.7   **B** 25.7, 49.5   **C** 37.6, 49.5   **D** 37.6, 75.2

20) In Fig. ST6.3 then, correct to 3 s.f., the area of triangle OPQ is

    A 569 mm^2 **B** 669 mm^2 **C** 696 mm^2 **D** 966 mm^2

21) Correct to 3 s.f., the distance AB in Fig. ST6.4 is

    A −13.0   **B** −12.9   **C** 12.9   **D** 13.0

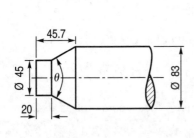

Fig. ST6.2

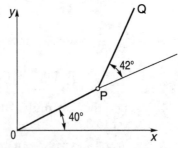

Fig. ST6.3

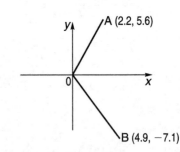

Fig. ST6.4

22) The sine rule states $\dfrac{a}{\sin A} = \dfrac{b}{\sin B} = \dfrac{c}{\sin C}$ and if we tranpose the equation for C we get

(i) $\dfrac{c}{b} \sin B$ or $\dfrac{c}{a} \sin A$

(ii) inv sin $\left(\dfrac{c}{b} \sin B\right)$ or inv sin $\left(\dfrac{c}{a} \sin A\right)$

Decide whether each of these statements is true or false

    A (i) is true and (ii) is true   **B** (i) is true and (ii) is false
    C (i) is false and (ii) is true   **D** (i) is false and (ii) is false

Answers to exercises

Exercise 1.1
1) 8 km   2) 15 Mg
3) 3.8 Mm   4) 1.8 Gg
5) 7 mm   6) 1.3 μm
7) 28 g   8) 360 mm
9) 64 mg   10) 3.6 mA

Exercise 1.2
1) 0.45 V
2) (a) 2 kN   (b) 4 ms^{-2}
3) 30 kJ   4) 60 kW
5) 2 kW   6) 0.111 MJ
7) 147 kJ, 4.9 kW

Exercise 1.3
1) 805   2) 54.9
3) 0.391   4) 0.227
5) 44.0   6) 1270
7) 49.2
8) 23.01 ± 0.03 mm
9) 0.110, 0.020 9, 0.000 492
10) (a) 24.1   (b) 683
(c) 0.683
11) 0.011
12) (a) 88.0   (b) 26.8
(c) 96.6
13) 42.5

Exercise 1.4
1) 24.865 8, 24.87, 25
2) 0.008 357, 0.008 36, 0.008 4
3) 4.978 5, 4.98, 5
4) 22   5) 35.60
6) 28 388 000, 28 000 000
7) 4.149 8, 4.150, 4.15
8) 9.20
9) (a) 2.138 9   (b) 2.139
(c) 2.14
10) (a) 25.17   (b) 25.2
11) (a) 0.003 99   (b) 0.004 0
(c) 0.004
12) (a) 7.204   (b) 7.20
(c) 7.2
13) (a) 0.726   (b) 0.73

Exercise 1.5
1) 64.5, 63.5, 0.781%
2) 2474, 2464, 0.203%
3) 3.075, 3.065, 0.163%
4) 0.65, 0.55, 8.33%

5) 29.95, 28.85, 0.55
6) 1.315, 1.205, 0.055
7) 2.80, 2.60, 0.1
8) 0.7515, 0.7405, 0.0055
9) 39.07 ± 0.005, 0.005, 0.013%
10) 0.372 ± 0.0005, 1.238 ± 0.0005, 3.222, 3.218, 0.002, 0.0621%
11) 0.553, 0.844%
12) 7.00 ± 0.005, 0.005, 0.0714%
13) 12.015, 11.985, 0.015, 0.125%

Exercise 1.6
1) 13.1   2) −11.35
3) 27.4   4) 0.001 49
5) 1.94   6) −4.26
7) 1.28   8) 18.8
9) −2.52   10) 527
11) −22.8   12) −22.8
13) 0.007 6   14) −0.348
15) 0.66   16) 0.55
17) −4.1   18) 4.0
19) 6.6   20) 0.001 30

Exercise 1.7
1) 29.0   2) 12.3
3) 0.0391   4) 0.0160
5) 0.0103   6) 94.2
7) 42.1   8) 86.3
9) 0.67   10) 2.90
11) 2.54   12) 0.51
13) 17.11   14) 2.84
15) 1.62   16) 0.52
17) 470   18) 3.22
19) 63   20) 1.92×10^8
21) 1.2   22) 9.2
23) 70.3   24) 55.8°
25) (a) 72.4   (b) 3.22
(c) 244   (d) 10.7

Exercise 1.8
1) (a) 51   (b) 143
(c) 307
2) (a) 35_8 (b) 363_8
(c) 3404_8
3) (a) 715   (b) 5614
5) (a) 10111   (b) 1111101
(c) 1100001
6) (a) 22   (b) 57
(c) 90

7) (a) 0.8125   (b) 0.4375
(c) 0.1875
8) (a) 0.011   (b) 0.0101
(c) 0.111
9) (a) 0.0010101
(b) 10010.0111011
(c) 1101100.1011010
10) (a) 1110   (b) 100110
(c) 111000
11) (a) 100011   (b) 1101110
(c) 101011111

Exercise 1.9
1) 8.8 mm   2) 0.0128 m^2
3) (a) 1200 mm^2
(b) 276 mm^2 (c) 261 mm^2
(d) 774 mm^2 (e) 1050 mm^2
4) (a) 892 mm^2 (b) 3060 mm^2
5) (a) 1380 mm^2 (b) 6500 mm^2
(c) 331 mm^2 (d) 1930 mm^2
6) 29.9 mm   7) 302 mm^2
8) 34.1 mm^2 9) 2592 mm^2
10) 909   11) 1910 mm

Exercise 1.10
1) 335 mm
2) 0.008 75 m^3
3) 477 mm   4) 1.51 m^2
5) 19.9 ℓ 6) 75.4 mm
7) 5.33 m^3, 20.5 m^2
8) 0.437 m^3
9) (a) 0.366 m^3 (b) 0.583 m
10) 2.09 ℓ
11) 14 m^3, 22.7 m^2
12) 150 mm

Exercise 2.1
1) 2^{11} 2) a^8 3) n^3
4) 3^{11} 5) b^{-3} 6) 10^4
7) z^3 8) 3^{-4} 9) m^4
10) x^{-3} 11) 9^{12} 12) y^{-6}
13) t^8 14) c^{14} 15) a^{-9}
16) 7^{-12} 17) b^{10} 18) s^{-9}
19) 8   20) 1   21) 0.5
22) 8   23) 0.25   24) 100
25) 0.25   26) 0.143   27) 0.04
28) 3   29) 7   30) 25
31) 7   32) 3.375   33) 0.003 91

Exercise 2.2

1) 5 2) 2 3) 2
4) 2 5) 4 6) 125
7) 8 8) 27 9) 2
10) 4 11) 0.167 12) 0.125
13) 0.031 25 14) 64
15) 3

Exercise 2.3

1) $x^{1/2}$ 2) $x^{4/5}$ 3) $x^{-1/2}$
4) $x^{-1/3}$ 5) $x^{-4/3}$ 6) $x^{-3/2}$
7) $x^{2/3}$ 8) $x^{0.075}$ 9) $x^{2/3}$
10) $x^{1/3}$ 11) $x^{1/3}$ 12) x
13) $a^{-13/6}$ 14) $a^{-11/3}$ 15) $x^{4.5}$
16) $b^{1/2}$ 17) $m^{7/4}$ 18) $z^{2.3}$
19) 1 20) $u^{-5/2}$ 21) $y^{1/4}$
22) $n^{1/4}$ 23) $x^{11/14}$ 24) $t^{-2/3}$

Exercise 2.4

1) $\log_a n = x$ 2) $\log_2 8 = 3$
3) $\log_5 0.04 = -2$
4) $\log_{10} 0.001 = -3$
5) $\log_x 1 = 0$
6) $\log_{10} 10 = 1$
7) $\log_a a = 1$
8) $\log_e 7.39 = 2$
9) $\log_{10} 1 = 0$ 10) 3
11) 3 12) 4 13) 3
14) 9 15) 64 16) 100
17) 1 18) 2 19) 3
20) $\frac{1}{2}$ 21) 1

Exercise 2.5

1) (a) $2\log a + \log b$
   (b) $\log a + 3\log c - 4\log b$
   (c) $\log a + \log b - \log c - \log d$
2) (a) 1.324 (b) −1.709
   (c) 1.544 (d) −1.108
3) (a) 15.80 (b) 1.094
   (c) 0.031 43 (d) 0.581 0
4) (a) 85.1 (b) 1.98
   (c) 0.987 (d) 0.265
5) 2.485 6) 0.096 91
7) 3.466

Exercise 2.6

1) 2.08 2) 368
3) 0.795 4) 0.000 179
5) 0.267 6) 1.77
7) −7.38 8) −0.322
9) 4.63 10) 2.35
11) −2.28 12) −36.5
13) 22.2

Exercise 2.7

1) (a) 853 (b) 39.3
   (c) 2.18
2) 0.003 57 3) 0.0164
4) 0.590 5) 25.7
6) 132 7) 4.44
8) 44.2 9) 0.005 49
10) 0.769

Exercise 3.1

1) $9x + 6y$ 2) $10p - 15q$
3) $-a + 2b$ 4) $-4x - 12$
5) $2k^2 - 10k$ 6) $-9xy - 12y$
7) $ap - aq - ar$
8) $4abxy - 4acxy + 4dxy$
9) $3x^4 - 6x^3y + 3x^2y^2$
10) $-14P^3 + 7P^2 - 7P$
11) $2m - 6m^2 + 4mn$
12) $x + 7$ 13) $16 - 17x$
14) $7x - 11y$ 15) $\dfrac{7y}{6} - \dfrac{3}{2}$
16) $-8a - 11b + 11c$
17) $3a - 9b$
18) $-x^3 + 18x^2 - 9x - 15$

Exercise 3.2

1) $x^2 + 9x + 20$
2) $2x^2 + 11x + 15$
3) $6x^2 + 16x + 8$
4) $10x^2 + 17x + 3$
5) $21x^2 + 41x + 10$
6) $x^2 - 4x + 3$
7) $x^2 + 2x - 3$ 8) $x^2 + 5x - 14$
9) $x^2 - 2x - 15$ 10) $2x^2 + x - 10$
11) $6x^2 + x - 15$
12) $12x^2 + 4x - 21$
13) $2p^2 - 7pq + 3q^2$
14) $6v^2 - 5uv - 6u^2$
15) $6a^2 + ab - b^2$
16) $x^2 + 2x + 1$
17) $4x^2 + 12x + 9$
18) $9x^2 + 42x + 49$
19) $x^2 - 2x + 1$
20) $4x^2 - 12x + 9$
21) $x^2 + 2xy + y^2$
22) $P^2 + 6PQ + 9Q^2$
23) $9x^2 - 24xy + 16y^2$
24) $4x^2 - y^2$
25) $4m^2 - 9n^2$
26) $x^4 - y^2$
27) $x^3 + 2x^2 - 5x - 6$
28) $2x^3 - 3x^2 - 11x + 6$

29) $x^3 - 3x - 2$
30) $x^4 - 1$
31) $x^3 + (a + b + c)x^2$
    $+(ab + bc + ca)x + abc$

Exercise 3.3

1) p^2q 2) ab^2 3) $3mn$
4) b 5) $3xyz$ 6) $2(x + 3)$
7) $4(x - y)$ 8) $5(x - 1)$
9) $4x(1 - 2y)$ 10) $m(x - y)$
11) $x(a + b + c)$ 12) $\dfrac{1}{2}\left(x - \dfrac{y}{4}\right)$
13) $5(a - 2b + 3c)$ 14) $ax(x + 1)$
15) $\pi r(2r + h)$ 16) $3y(1 - 3y)$
17) $ab(b^2 - a)$
18) $xy(xy - a + by)$
19) $5x(x^2 - 2xy + 3y^2)$
20) $3xy(3x^2 - 2xy + y^4)$
21) $I_0(1 + at)$
22) $\dfrac{1}{3}\left(x - \dfrac{y}{2} + \dfrac{z}{3}\right)$
23) $a(2a - 3b) + b^2$
24) $x(x^2 - x + 7)$
25) $\dfrac{m^2}{pn}\left(1 - \dfrac{m}{n} + \dfrac{m^2}{pn}\right)$

Exercise 3.4

1) $(x + y)(a + b)$
2) $(p - q)(m + n)$
3) $(ac + d)^2$
4) $(2p + q)(r - 2s)$
5) $2(a - b)(2x + 3y)$
6) $(x^2 + y^2)(ab - cd)$
7) $(mn - pq)(3x - 1)$
8) $(k^2l - mn)(l - 1)$

Exercise 3.5

1) $(x + 1)(x + 3)$
2) $(x + 2)(x + 4)$
3) $(x - 1)(x - 2)$
4) $(x + 5)(x - 3)$
5) $(x + 7)(x - 1)$
6) $(x + 2)(x - 7)$
7) $(x + y)(x - 3y)$
8) $(2x + 3)(x + 5)$
9) $(p + 1)(3p - 2)$
10) $(2x + 1)(2x - 6)$
11) $(m + 2)(3m - 14)$
12) $(3x + 1)(7x + 10)$
13) $(2a + 5)(5a - 3)$
14) $(2x + 5)(3x - 7)$
15) $(2p + 3q)(3p - q)$

Exercise 3.5 (cont.)

16) $(4x + y)(3x - 2y)$
17) $(x + y)^2$ 18) $(2x + 3)^2$
19) $(p + 2q)^2$ 20) $(3x + 1)^2$
21) $(m - n)^2$ 22) $(5x - 2)^2$
23) $(x - 2)^2$
24) $(m + n)(m - n)$
25) $(2x + y)(2x - y)$
26) $(3p + 2q)(3p - 2q)$
27) $(x + 1/3)(x - 1/3)$
28) $(1 + b)(1 - b)$
29) $(1/x + 1/y)(1/x - 1/y)$
30) $(11p + 8q)(11p - 8q)$

Exercise 3.6

1) $6a^2$ 2) $2x^2y$ 3) m^2n^2
4) $2abc^2$ 5) $2(x + 1)$
6) $x^2(a + b)^2$
7) $(a + b)(a - b)$
8) $x(1 - x)(x + 1)$

Exercise 3.7

1) $\dfrac{1}{ab}$ 2) $\dfrac{a}{b}$ 3) $\dfrac{x^2}{y^2}$

4) $\dfrac{xy}{2}$ 5) $\dfrac{1}{ab}$ 6) $c(a + b)$

7) $1 - x^2$ 8) $\dfrac{c}{(a - b)^2}$ 9) $\dfrac{x + y}{xy}$

10) $\dfrac{a + 1}{a}$ 11) $\dfrac{m - n}{n}$ 12) $\dfrac{b - c^2}{c}$

13) $\dfrac{ad - bc}{bd}$ 14) $\dfrac{ac - 1}{bc}$

15) $\dfrac{1 + y + xy}{xy}$ 16) $\dfrac{12 + x^2}{4x}$

17) $\dfrac{3de + 2ce - 5cd}{cde}$

18) $\dfrac{ad + cb + bd}{bd}$

19) $\dfrac{2h - 5f - 3g}{6fgh}$ 20) $\dfrac{4 - 2x}{x(x + 2)}$

21) $\dfrac{2}{2 - x}$

Exercise 3.8

1) $1 + \dfrac{b}{a}$ 2) $\dfrac{1}{b} - \dfrac{1}{a}$ 3) $\dfrac{1}{c} + 1$

4) $\dfrac{x}{2} + \dfrac{y}{2x}$ 5) $\dfrac{a}{bc} - \dfrac{1}{c} + \dfrac{1}{b}$

6) $\dfrac{1}{(x - y)} + \dfrac{1}{x}$

Exercise 3.9

1) $\dfrac{x}{x + 1}$ 2) $(x^2 - 1)$

3) $\dfrac{1}{a^2 - 1}$ 4) $\dfrac{u}{1 - u}$

5) $x + y$ 6) $\dfrac{R_1 R_2}{R_1 + R_2}$

Exercise 4.1

1) $m = 2$ 2) $x = 5$
3) $m = 5$ 4) $x = -29/5$
5) $x = -2$ 6) $x = 45/8$
7) $x = -15$ 8) $m = 15/28$
9) $t = 6$ 10) $y = -70$
11) $x = 5/3$ 12) $x = 13$
13) $u = 9/7$ 14) $x = 3.5$
15) $v = 20$ 16) $x = 4$
17) $i_1 = 9/7$ A
18) $l = 48.75$ mm
19) $d = 300$ mm   20) $R_2 = 6\,\Omega$
21) $x = 4$ 22) $R_1 = 2.85\,\Omega$

Exercise 4.2

1) $x^2 - 4x + 3 = 0$
2) $x^2 + 2x - 8 = 0$
3) $x^2 + 3x + 2 = 0$
4) $x^2 - 2.3x + 1.12 = 0$
5) $x^2 - 1.07x - 4.53 = 0$
6) $x^2 + 7.32x + 12.19 = 0$
7) $x^2 - 1.4x = 0$
8) $x^2 + 4.36x = 0$
9) $x^2 - 12.25 = 0$
10) $x^2 - 8x + 16 = 0$

Exercise 4.3

1) ± 6 2) ± 1.25
3) ± 1.333 4) -4 or -5
5) 8 or -9 6) 2 or $\frac{1}{3}$
7) 3 8) 4 or -8
9) $\frac{4}{7}$ or $\frac{3}{2}$ 10) $\frac{7}{3}$ or $-\frac{4}{3}$
11) 1.175 or -0.425
12) $\frac{5}{6}$ or $\frac{1}{6}$
13) 0.573 or -2.907
14) 0.211 or -1.354
15) 1 or -0.2
16) 3.886 or -0.386
17) 0.956 or -1.256
18) 2.388 or 0.262
19) 0.44 or -3.775
20) 8.385 or -2.385
21) -0.225 or -1.775
22) 11.14 or -3.14
23) 1.303 or -2.303

24) ± 53.67
25) 5.24 or 0.76
26) -3.064 or -0.935

Exercise 4.4

1) 149.6 2) 92.4 mm
3) 65 mm $\times$ 95 mm
4) 40 mm
5) 0.685 or 23.3 mm
6) 30 or 72 mm   7) 54.6 mm
8) 50 9) 2.88 m
10) 94.6×94.6 mm
11) 2.41 and -0.41 s

Exercise 4.5

1) $1, 2$ 2) $4, 5$ 3) $4, 1$
4) $7, 3$ 5) $\frac{1}{2}, \frac{3}{4}$ 6) $7, 10$
7) $3, 2$ 8) $5, 2$

Exercise 4.6

1) $0.2, 1.3, 3.7$ 2) £224, £168
3) £1.60, £0.80   4) $0.3, 0.2, 4.7$
5) £0.50, £1.50
6) 9 and $7\,\text{g/cm}^3$
7) $19.0\,\text{m s}^{-1}, 3.11\,\text{m s}^{-2}$
8) $i_1 = 7.16, i_2 = 5.23$
9) $24\,\Omega, 0.00417$

Exercise 4.7

1) $x = 0.333,$ $y = -1.667;$
   $x = 2,$ $y = 5$
2) $x = -1,$ $y = 6;$
   $x = 0.375,$ $y = -0.875$
3) $x = 1.175,$ $y = 2.175;$
   $x = -0.425,$ $y = 0.575$
4) $x = -0.969,$ $y = 0.956;$
   $x = 0.579,$ $y = -1.256$
5) 12 or 14.7 mm
6) 2 m, 8 m

Exercise 4.8

1) $T = \dfrac{pV}{nR}$ 2) $h = \dfrac{Hr}{R}$

3) $u = v - at$ 4) $t = \dfrac{v - u}{a}$

5) $C = \frac{5}{9}(F - 32)$ 6) $x = \dfrac{y - c}{m}$

7) $r = 1 - \dfrac{a}{S}$ 8) $R = \dfrac{V}{I} - r$

9) $h = \dfrac{S}{\pi r} - r$ 10) $T = \dfrac{H}{ws} + t$

11) $u^2 = v^2 - 2as$

Exercise 4.8 (cont.)

12) (a) $S = \frac{n}{2}[2a + (n-1)d]$

 (b) $a = \frac{S}{n} - \frac{d}{2}(n-1)$

13) $R_1 = \frac{R_2 R}{R_2 - R}$ 14) $R_2 = \frac{R_1 R}{R_1 - R}$

Exercise 4.9

1) $h = \frac{v^2}{2g}$ 2) $r = \sqrt{\frac{A}{\pi}}$

3) $v = \sqrt{\frac{2E}{m}}$ 4) $A = \frac{\pi d^2}{4}$

5) $f = \sqrt{\frac{2EU}{V}}$

6) $c = \sqrt{a^2 - b^2}$ 7) $l = \frac{g}{4f^2\pi^2}$

8) $M = \sqrt{T_e^2 - T^2}$

9) $C = \frac{\pi}{L}\sqrt{\frac{EI}{P}}$

10) $v = \sqrt{\frac{2}{m}(E_t - mgh)}$

11) $b = \sqrt{12k^2 - a^2}$

12) $L = \sqrt{12k^2 - 3R^2}$

13) $f = \frac{(D^2 + d^2)p}{D^2 - d^2}$

14) $P = \sqrt{4Q_e^2 - Q^2}$

15) $Q = \frac{1}{2}\sqrt{(2P_e - P)^2 - P^2}$

Exercise 5.1

1) (0, 50.00), (47.55, 15.45),
 (29.39, −40.45),
 (−29.39, −40.45),
 (−47.55, 15.45)

2) (50, 90°), (50, 18°), (50, −54°),
 (50, −126°), (50, 162°)

3) (34.64, 20.00), (20.00, −34.64),
 (−34.64, −20.00),
 (−20.00, 34.64)

4) (40, 30°), (40, −60°),
 (40, −150°), (40, 120°)

5) (35.24, 12.83), (5.22, −37.14),
 (−18.75, 32.48)

6) (37.50, 20°), (37.50, −82°),
 (37.50, 120°)

7) (36.94, 47.28), (−19.53, 56.73)

8) (60, 52°), (60, 109°)

9) (14.28, 8.57), (35.72, 21.43)

10) (16.65, 30.96°), (41.65, 30.96°)

Exercise 5.2

1) 1.897 2) 3.0248
3) −0.28 4) $-\frac{4}{5}$, $-\frac{3}{4}$
5) 8°14′, 171°47′
6) 153°13′, 206°47′
7) (a) 45°32′, 134°28′
 (b) 118°46′ (c) 43°28′
 (d) 119°33′
8) (a) 232° and 308°
 (b) 304° (c) 289°
9) 14°34′, 165°26′

Exercise 5.3

1) (a) 0.6109 (b) 1.457
 (c) 0.3367 (d) 0.7621
2) (a) 9°55′25″ (b) 89°33′53″
 (c) 4°29′11″
3) (a) 1.05 m (b) 22.9 mm
4) (a) 120° (b) 10.2°
5) 57.3 mm, 2005 mm²
6) 866 mm²
7) (a) 29.3 mm, 80.7 mm
 (b) 104 mm²
8) 163 mm²
9) 369 mm, 20 600 mm²

Exercise 6.1

1) 38.39 mm 2) 32.5 mm
3) 20.4 mm 4) 37.9 mm

Exercise 6.2

1) 24.3 2) 58.2
3) 23.1 4) 121
5) 39.7 6) 255
7) 30°33′ 8) 21°33′
9) 64°45′ 10) 177 mm

Exercise 6.3

1) 45.79 mm 2) 19.95 mm
3) 20.90 mm 4) 24.98 mm
5) 10°44′

Exercise 6.4

1) 1.64 mm
2) 1°31′; 13.04 mm; 9.59 mm
3) 65°46′; 29.71 mm
4) 53.01 mm 5) 31.99 mm
6) 4°24′; 25.51 mm
7) 104.98 mm 8) 5.18 mm
9) 2.887 mm; 53.44 mm
10) 30.53 mm

Exercise 6.5

1) 2408 mm 2) 5369 mm
3) 12.63 m 4) 1287 mm
5) 2971 mm 6) 5740 mm
7) 16.01 m 8) 2215 mm
9) 36.1 m
10) BD = DF = AC = CE = 2.66 m,
 GB = FH = 4.41 m,
 BC = CF = 3.59 m,
 GA = EH = 3.70 m

Exercise 7.1

1) (a) C = 71°, b = 59.1 mm
 c = 99.9 mm
 (b) A = 48°, a = 71.5 mm
 c = 84.2 mm
2) (a) c = 10.2 mm
 A = 50°11′,
 B = 69°49′,
 (b) a = 11.8 m,
 B = 44°42′,
 C = 79°18′
3) 64.00 mm 4) 37°35′
5) 40.5° 6) 41°27′
7) (a) 14.2 m (b) 142°39′
9) 21.2 A 10) 13.4, 14°52′
11) 18.6 A

Exercise 8.1

1) 0.121 m²
2) 2765 mm each side
3) 540 m² 4) 738 mm²
5) 0.420 kg 6) 11.7 mm

Exercise 9.1

1)

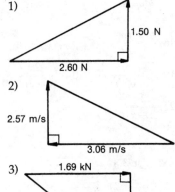

2)

3)

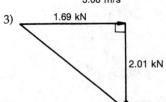

Exercise 9.1 (cont.)

4)

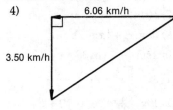

6.06 km/h

3.50 km/h

5)

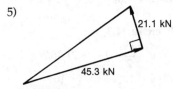

21.1 kN

45.3 kN

6) 5.00 kN down the slope,
   8.66 kN at right angles to slope
7) 81.6 N horizontal,
   40.1 N vertical
8) 1.93 m/s   9) 58.3 N, 31°
10) 703.5 km/h, 5.71° east of north
11) 0.568 m/s, 38.4° west of north
12) 353 N, anticlockwise 11.0°
13) 347 N along the track
    67.4 N between wheel flanges
    and rail
14) 3.61 V, anticlockwise 33.7°
15) 4.78 A, clockwise 13.9°
16) 10.1 V, clockwise 40.8°
17) 6.51 A, clockwise 33.0°
18) F_{AB} 76 kN tensile
19) F_{CB} 143 kN compressive

Exercise 10.1

1) $m = 1, c = 3$
2) (a) $m = 1, c = 3$
   (b) $m = -3, c = 4$
   (c) $m = -3.1, c = -1.7$
   (d) $m = 4.3, c = -2.5$
   (e) infinite, none
   (f) zero, 2.9
   (g) $m = 1, c = -4$
   (h) $m = -0.5, c = 1.5$
   (j) $m = 0.556, c = $ zero

Exercise 10.2

1) $m = 2, c = 1$
2) $a = 0.25, b = 1.25$
3) $a = 0.29, b = -1.0$
4) 529 N

Exercise 10.3

1) $E = 0.0984 W + 0.72$
2) $a = 0.03, b = 0$

3) $a = 100, b = 0.43$
4) 524 N/m   5) 51 ohms

Exercise 10.4

1) b   2) d
3) e   4) a
5) c   6) b

Exercise 10.5

1) $y = 1.82, x = 0.84$
2) $x = -1.39, -0.5$
3) (a) $T = 631$ (b) $s = 1.12$
4) 4.34 mA per second
5) (a) 0.18 seconds
   (b) 1200 volts per second

Exercise 10.6

1) 1, 2   2) 4, 5
3) 18, 28
4) 5400, 1700, £5400, £3400
5) 25 ohm, 0.005, 31.3 ohm
6) £250, £6500
7) £14 000, £16 000
8) 0.4, 50, 170 N

Exercise 10.7

1) 3, 4   2) 4 repeated
3) +3, −3   4) 3, −5
5) 0.667, 7
6) $r = 22.5$ mm
7) $v_1 = 0.495$ V

Exercise 10.8

1) $x = -1$ $x = 4$
   $y = 1$ ' $y = 6$
2) $x = 0$ $x = 3$
   $y = 5$ ' $y = 11$
3) $x = 1$ $x = -0.2$
   $y = 3$ ' $y = -3$
4) $x = 2.39$ $x = 0.26$
   $y = 6.91$ ' $y = 0.54$
5) 60 mm × 80 mm

Exercise 11.1

5) (a) $\sin\left(\omega t + \dfrac{\pi}{2}\right)$
   (b) $\sin(\omega t - \pi)$
   (c) $\sin \omega t$
   (d) $\sin\left(\omega t + \dfrac{\pi}{6}\right)$

Exercise 12.1

1) $a = 70, b = 50$
2) $k_m = 0.016 + \dfrac{0.023}{\mu}$
3) Gradient = 1500,
   Intercept = 0
4) $k = 0.2$
5) $a = 0.761, b = 10.1$
6) $m = 0.040, c = 0.20$
7) $m = 0.1, c = 1.4$

Exercise 12.2

1) $a = 3, n = 0.5$
2) $n = 4.05$, for $I = 20$
         read $I = 35$
3) $t = 0.3\, m^{1.5}$
4) $k = 100, n = -1.2$
5) $a = 245, b = 33$
6) $\mu = 0.5, k = 5$
7) $I = 0.02, T = 0.2$
8) $k = 23.3, c = 2.99$
9) $V = 100, t = 0.0025$

Exercise 13.1

1) −4, 8   2) 3   3) 2

Exercise 13.2

1) $2x$ 2) $7x^6$ 3) $12x^2$
4) $30x^4$ 5) $1.5t^2$ 6) $2\pi R$
7) $\frac{1}{2}x^{-1/2}$ 8) $6x^{1/2}$ 9) $x^{-1/2}$
10) $2x^{-1/3}$ 11) $-2x^{-3}$
12) $-x^{-2}$ 13) $-\frac{3}{5}x^{-2}$
14) $-6x^{-4}$ 15) $-\frac{1}{2}x^{-3/2}$
16) $-\frac{1}{3}x^{-3/2}$ 17) $-\frac{15}{2}x^{-5/2}$
18) $\frac{3}{10}t^{-1/2}$ 19) $-0.01h^{-2}$
20) $-35x^{-8}$ 21) $8x - 3$
22) $9t^2 - 4t + 5$ 23) $4u - 1$
24) $20x^3 - 21x^2 + 6x - 2$
25) $35t^4 - 6t$
26) $\frac{1}{2}x^{-1/2} + \frac{5}{2}x^{-3/2}$
27) $-3x^{-2} + 1$
28) $\frac{1}{2}x^{-1/2} - \frac{1}{2}x^{-3/2}$
29) $3x^2 - \frac{3}{2}x^{-3/2}$
30) $1.3t^{0.3} + 0.575t^{-3.3}$
31) $\frac{9}{5}x^2 - \frac{4}{7}x + \frac{1}{2}x^{-1/2}$
32) $-0.01x^{-2}$
33) $4.65x^{0.5} - 1.44x^{-0.4}$
34) $\frac{3}{2}x^2 + 5x^{-2}$
35) $-6 + 14t - 6t^2$
36) −5, 19   37) 39.5, 5, 17
38) 2.5, 2, 1

Exercise 13.3

1) $2\cos 2\theta - 5\sin 5\theta$

2) $4\cos(4t + \pi)$

3) $-7\sin\left(7t - \dfrac{3\pi}{2}\right)$

4) $\dfrac{1}{x}$

5) $-\dfrac{1}{x}$

6) $\dfrac{2}{x}$

7) $3e^{3u} + 3e^{-3u}$ 8) $-30e^{-5v}$

Exercise 13.4

1) $42\,\text{m/s}$ 2) $-6\,\text{m/s}^2$

3) (a) $6\,\text{m/s}$

 (b) 2.41 or $-0.41\,\text{s}$

 (c) $6\,\text{m/s}^2$ (d) $1\,\text{s}$

4) $-0.074\,\text{m/s}$, $0.074\,\text{m/s}^2$

5) $10\,\text{m/s}$, $30\,\text{m/s}$

6) $3.46\,\text{m/s}$

7) (a) $4\,\text{rad/s}$ (b) $36\,\text{rad/s}^2$

 (c) $0\,\text{s}$ or $1\,\text{s}$

8) (a) $-2.97\,\text{rad/s}$

 (b) $0.280\,\text{s}$

 (c) $-8.98\,\text{rad/s}^2$

 (d) $1.57\,\text{s}$

9) $62.5\,\text{kJ}$

Exercise 13.5

1) (a) 11 (max), -16 (min)

 (b) 4 (max), 0 (min)

 (c) 0 (min), 32 (max)

2) (a) 54   (b) 2.5

 (c) $x = -2$

3) $(3, -15)$, $(-1, 17)$

4) (a) -2 (b) 1

 (c) 9

5) (a) 12   (b) 12.48

6) $10\,\text{V (max)}$, $4\,\text{V (min)}$

7) 8

8) $12.5\,\text{m/s}$, $1.23\,\text{kW}$

9) $15\,\text{mm}$ 10) $10\,\text{m}$

11) 4

12) $108\,000\,\text{mm}^3$

13) $86\,\text{mm diam.}$, $86\,\text{mm long}$

14) $405\,\text{mm}$

15) $28.9\,\text{mm diam.}$, $14.4\,\text{mm high}$

Exercise 14.1

1) $353\,\text{m}$, $170\,\text{m}$ 2) $2.55\,\text{A}$

3) $85.6\,\text{kJ}$ 4) $1090\,\text{m}^2$

5) $17.2\,\text{m}^3$

Exercise 14.2

1) $\frac{1}{3}x^3 + c$ 2) $-\cos\theta + c$

3) $\frac{2}{3}x^{3/2} + c$ 4) $-\frac{1}{x} + c$

5) $(\log_e x) + c$ 6) $\frac{1}{8}e^{8t} + c$

7) $\frac{3}{2}(\sin 2\theta) + c$ 8) $\frac{5}{9}x^9 + e^x + c$

9) $(\log_e x) + \frac{1}{2}x^2 + 3x + c$

10) $2\sin\theta + \frac{1}{3}\cos 3\theta + c$

11) $6x + \dfrac{5}{2}x^2 + 2x^{1/2} - \dfrac{2}{x} + c$

12) $10u + \frac{1}{10}(e^{5u} - e^{-5u}) + c$

Exercise 14.3

1) $y = \dfrac{x^2}{2} + 1$ 2) 46.7

3) $y = 10 + 3x - x^2$, 10

4) $y = \frac{1}{3}e^{3t} - 2.70$

5) $p = \log_e t + 2.31$

6) $y = x^2 - 2x + 4$

7) $y = \sin x + 1.16$

8) $y = \sin\theta$

9) $y = 4.72 - 3\cos t$

Exercise 14.4

1) 2.33   2) 8   3) 0.511

4) 0   5) 30.6   6) 1.89

Exercise 14.5

1) $19.2\,\text{J}$ 2) $16.7\,\text{mm}$

3) $20.1\,\text{J}$ 4) $2.55\,\text{A}$

5) $85.6\,\text{kJ}$

Answers to sample unit tests

Sample unit test 1

1) A   2) A   3) B   4) C
5) C   6) B   7) B   8) D
9) A   10) D   11) A   12) D
13) C   14) C   15) B   16) D
17) C   18) C   19) D   20) A
21) A   22) D

Sample unit test 2

1) B   2) B   3) B   4) A
5) B   6) A   7) A   8) C
9) C   10) C   11) D   12) B
13) C   14) A   15) B   16) B
17) C   18) C   19) A   20) B
21) C   22) C

Sample unit test 3

1) C   2) B   3) A   4) B
5) B   6) D   7) D   8) A
9) D   10) C   11) D   12) D
13) B   14) D   15) C   16) C
17) C   18) B   19) A   20) D
21) D   22) D

Sample unit test 4

1) D   2) A   3) B   4) B
5) B   6) A   7) D   8) C
9) C   10) A   11) D   12) B
13) B   14) C   15) B   16) C
17) D   18) C   19) C   20) A
21) D   22) B

Sample unit test 5

1) A   2) C   3) C   4) C
5) D   6) C   7) B   8) A
9) C   10) D   11) C   12) D
13) A   14) C   15) C   16) A
17) B   18) D   19) B   20) B
21) C   22) A

Sample unit test 6

1) C   2) A   3) C   4) A
5) D   6) D   7) D   8) D
9) D   10) D   11) B   12) C
13) D   14) C   15) A   16) C
17) B   18) B   19) D   20) B
21) D   22) C

Index